GRADE 6 MATH
WORKBOOK

with ANSWERS

order of operations

prime factorization

fractions/decimals

geometric figures

financial math

ratios/proportions

prealgebra skills

data analysis

histograms

exponents

$$\frac{8}{?} = \frac{48}{72}$$

$$(3+9)^2 \div 8 - 5 \times 3$$

Chris McMullen, Ph.D.

Grade 6 Math Workbook with Answers

Chris McMullen, Ph.D.

Zishka Publishing

ISBN: 978-1-941691-56-4

Mathematics > Arithmetic

Study Guides > Workbooks> Math

Education > Math > Grade 6

CONTENTS

Introduction 4

1 Arithmetic Operations 5

2 Fractions 31

3 Decimals and Percents 55

4 Proportions 75

5 Variables 101

6 Relationships 127

7 Data Analysis 147

8 Coordinate Graphs 173

9 Geometry 197

10 Finance 225

Answer Key 241

Glossary 305

Index 321

INTRODUCTION

This workbook covers a variety of topics that are typically taught in sixth grade mathematics. Each section begins by reviewing essential terms, concepts, and problem-solving strategies. The techniques are illustrated in examples that should serve as a helpful guide for completing the exercises at the end of the section. The answer to every problem can be found in the Answer Key at the back of the book. Topics include:

- order of operations
- fractions, decimals, and percents
- prime factorization
- ratios and rates
- prealgebra skills
- interpreting histograms
- geometric figures
- the coordinate plane
- and much more

-1-

ARITHMETIC OPERATIONS

1.1 Quick review of arithmetic facts

1.2 Negative values on a number line

1.3 Arithmetic with negative numbers

1.4 Whole number exponents

1.5 Rounding whole numbers

1.6 Comparing integers

1.7 Multiplication and division operators

1.8 Properties of arithmetic operators

1.9 Order of operations

1.10 The distributive property

1.11 Divisibility tests

1.12 Prime factorization

1.13 Greatest common factor

1.14 Factoring out the GCF

1.15 Least common multiple

1.16 Square roots

1.17 Factor out perfect squares

1.1 Quick review of arithmetic facts

Fluency with arithmetic facts is essential in math.

Problems. Practice these addition facts.

① $6 + 7 = 13$ 　② $8 + 7 = 15$ 　③ $9 + 8 = 17$

④ $7 + 9 = 16$ 　⑤ $8 + 8 = 16$ 　⑥ $5 + 9 = 14$

⑦ $7 + 7 = 14$ 　⑧ $9 + 6 = 15$ 　⑨ $8 + 5 = 13$

⑩ $8 + 6 = 14$ 　⑪ $7 + 5 = 12$ 　⑫ $9 + 9 = 18$

Problems. Practice these subtraction facts.

⑬ $16 - 7 = 9$ 　⑭ $14 - 9 = 5$ 　⑮ $15 - 7 = 8$

⑯ $15 - 8 = 7$ 　⑰ $15 - 6 = 9$ 　⑱ $16 - 9 = 7$

⑲ $13 - 6 = 7$ 　⑳ $17 - 9 = 8$ 　㉑ $14 - 6 = 8$

㉒ $16 - 8 = 8$ 　㉓ $14 - 7 = 7$ 　㉔ $17 - 8 = 9$

Problems. Practice these multiplication facts.

① $5 \times 5 = 25$ ② $8 \times 6 = 48$ ③ $9 \times 5 = 45$

④ $7 \times 5 = 35$ ⑤ $9 \times 7 = 63$ ⑥ $7 \times 7 = 49$

⑦ $6 \times 6 = 36$ ⑧ $5 \times 8 = 40$ ⑨ $6 \times 9 = 54$

⑩ $7 \times 8 = 56$ ⑪ $9 \times 9 = 81$ ⑫ $5 \times 6 = 30$

⑬ $9 \times 8 = 72$ ⑭ $6 \times 7 = 42$ ⑮ $8 \times 8 = 64$

Problems. Practice these division facts.

⑯ $27 \div 3 = 9$ ⑰ $56 \div 7 = 8$ ⑱ $45 \div 5 = 9$

⑲ $54 \div 9 = 6$ ⑳ $36 \div 6 = 6$ ㉑ $48 \div 8 = 6$

㉒ $49 \div 7 = 7$ ㉓ $30 \div 5 = 6$ ㉔ $63 \div 9 = 7$

㉕ $28 \div 4 = 7$ ㉖ $81 \div 9 = 9$ ㉗ $36 \div 4 = 8$

㉘ $72 \div 8 = 9$ ㉙ $42 \div 6 = 7$ ㉚ $64 \div 8 = 8$

1.2 Negative values on a number line

A negative number is the opposite of a positive number.

- Walking 50 yards north is the opposite of walking 50 yards south. We can call these 50 and -50.

- If you receive a paycheck of $400 and you spend $80, we can think of the expense as being $-$80.

On a number line, negative values are left of the zero line.

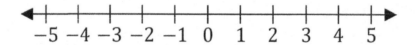

Example. Draw and label -7 on the number line below.

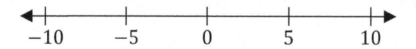

-7 is closer to 0 than -10 and further from 0 than -5.

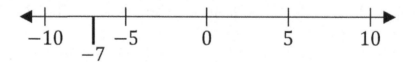

Problems. Draw and label each value on the number line.

① -3 ② 6   ③ -8 ④ -1

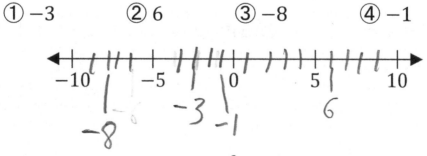

1.3 Arithmetic with negative numbers

Rule	Example
$a + (-b) = a - b$	$5 + (-3) = 5 - 3 = 2$
$-a + b = b - a$	$-5 + 3 = 3 - 5 = -2$
$-a + (-b) = -(a + b)$	$-5 + (-3) = -(5 + 3) = -8$
$a - (-b) = a + b$	$5 - (-3) = 5 + 3 = 8$
$-a - b = -(a + b)$	$-5 - 3 = -(5 + 3) = -8$
$-a - (-b) = -(a - b)$	$-5 - (-3) = -(5 - 3) = -2$
$-(-a) = a$	$-(-5) = 5$

Try to think your way through the rules so that you don't need to memorize them. For example, $4 - (-2) = 4 + 2 = 6$ because two minus signs make a plus sign, and $-4 - 2 = -6$ because both minus signs work together whereas $2 - 4 = -2$ because the signs oppose each other (and because -4 is more negative than 2 is positive).

Problems. Add or subtract negative numbers.

① $9 + (-2) =$ 7

② $7 - (-4) =$ 11

③ $-6 + (-3) =$ -3

④ $-2 - (-1) =$ -1

⑤ $4 + (-9) =$ -9

⑥ $-3 - 8 =$ -11

Rule	Example
$a \times (-b) = -a \times b$	$3 \times (-4) = -3 \times 4 = -12$
$-a \times (-b) = a \times b$	$-3 \times (-4) = 12$
$a \div (-b) = -a \div b$	$12 \div (-4) = -12 \div 4 = -3$
$-a \div (-b) = a \div b$	$-12 \div (-4) = 3$

When multiplying or dividing, if one of the given numbers is negative the answer is negative, but if both of the given numbers are negative the answer is positive.

Problems. Perform arithmetic with negative numbers.

① $5 \times (-3) =$ ② $18 \div (-6) =$

③ $-8 \times (-2) =$ ④ $-24 \div (-4) =$

⑤ $-4 \times 7 =$ ⑥ $-30 \div 5 =$

⑦ $-6 + 9 =$ ⑧ $-5 - (-7) =$

⑨ $-3 + (-3) =$ ⑩ $-4 - 4 =$

⑪ $-1 \times (-1) =$ ⑫ $-6 \div 6 =$

1.4 Whole number exponents

In the expression b^p, the base b is raised to the power of p. The power p is called an exponent. For an exponent that is a whole number, this means to multiply the base b by itself p times. When the power equals 2, it is called a square. For example, 8^2 reads as "8 squared" and equals $8 \times 8 = 64$. A power of 3 is called a cube. For example, 7^3 is "7 cubed." If the exponent is 1, the answer equals the base: $b^1 = b$. If the exponent is 0, the answer equals 1 regardless of the base: $b^0 = 1$. For example, $7^0 = 1$ and $7^1 = 7$.

Examples. (A) $2^4 = 2 \times 2 \times 2 \times 2 = 16$ (B) $9^0 = 1$ (C) $2^1 = 2$
(D) $5^2 = 5 \times 5 = 25$ (E) $10^3 = 10 \times 10 \times 10 = 1000$

Problems. Evaluate the following exponents.

① $2^3 = $ 8

② $3^2 = $ 9

③ $1^9 = $ 1

④ $12^1 = $ 12

⑤ $6^3 = $ 216

⑥ $20^0 = $ 00

⑦ $(-4)^3 = $

⑧ $10^6 = $ 1,000,000

⑨ $(-3)^4 = $

1.5 Rounding whole numbers

When rounding to the nearest ten, look at the units digit:

- If the units digit is 0 to 4, don't change the tens digit.

- If the units digit is 5 to 9, increase the tens digit by 1.

Either way, the units digit becomes 0.

Examples. Round each number to the nearest ten.

(A) 43 rounds to 40 because 3 is less than 5.

(B) 47 rounds to 50 because 7 is 5 or more.

(C) 45 rounds to 50 because 5 is 5 or more.

When rounding to the nearest 100, look at the tens digit.

Examples. Round each number to the nearest hundred.

(A) 349 rounds to 300 because 4 is less than 5.

(B) 350 rounds to 400 because 5 is 5 or more.

Problems. Round each number to the nearest ten.

① 22 ≈   ② 17 ≈   ③ 65 ≈

④ 96 ≈   ⑤ 5 ≈   ⑥ 4 ≈

Problems. Round each number to the nearest hundred.

⑦ 176 ≈   ⑧ 333 ≈   ⑨ 972 ≈

1.6 Comparing integers

To find the absolute value of a negative number, remove the minus sign. For example, the absolute value of -6 equals 6. The notation $|-6|$ reads as "the absolute value of -6." For a positive number, taking the absolute value has no effect.

Examples. (A) $|-3| = 3$ (B) $|0| = 0$ (C) $|-1| = 1$ (D) $|8| = 8$

Problems. Find the absolute values.

① $|-5| =$ ② $|-24| =$ ③ $|72| =$

If two numbers are both negative, the number with the smaller absolute value is larger. For example, -4 is larger than -7 because $|-4| = 4$ is smaller than $|-7| = 7$. Recall the greater than (>) and less than (<) signs: $-4 > -7$.

Examples. (A) $-2 > -5$ (B) $-9 < -3$ (C) $-4 < 4$ (D) $6 = 6$

(E) Order $-3, 1, -7$ from least to greatest: $-7, -3, 1$.

Problems. Use $>$, $<$, or $=$ to compare the given numbers.

④ -6 -8 ⑤ -9 2 ⑥ -8 -9

Order the given numbers from least to greatest.

⑦ $2, -1, -3$ ⑧ $-29, -31, -20$

1.7 Multiplication and division operators

In addition to the times symbol (×), multiplication is often represented by a dot (·) or using parentheses similar to the examples below.

$$6 \times 4 = 24 \qquad 6 \cdot 4 = 24 \qquad (6)(4) = 24 \qquad 6(4) = 24$$

Similarly, division is often represented by a slash (/) or a fraction.

$$18 \div 9 = 2 \qquad 18/9 = 2 \qquad \frac{18}{9} = 2$$

If the symbol x is used to represent an unknown quantity in an equation, it would be confusing to use the times (×) symbol for multiplication. It is conventional to simply put a number beside a variable. For example, $3x$ means 3 times the variable x.

Problems. Answer these arithmetic problems.

① $7 \cdot 3 =$ ② $(8)(2) =$ ③ $\frac{40}{5} =$

④ $(-4)(4) =$ ⑤ $\frac{28}{7} =$ ⑥ $5 \cdot 5 =$

⑦ $\frac{36}{4} =$ ⑧ $8 \cdot 7 =$ ⑨ $-3(-8) =$

1.8 Properties of arithmetic operators

Addition and multiplication are commutative because the order of the numbers doesn't matter. In contrast, division and subtraction aren't commutative.

$$a + b = b + a \qquad a \times b = b \times a$$

The associative property states that when three numbers are added or multiplied, the grouping doesn't matter.

$$a + (b + c) = (a + b) + c \qquad a \times (b \times c) = (a \times b) \times c$$

The distributive property can be expressed as follows. Note that it is often written without the times (×) symbol.

$$a \times (b + c) = a \times b + a \times c$$

The identity properties of addition and multiplication state that adding zero or multiplying by one have no effect.

$$a + 0 = a \qquad a \times 1 = a$$

Problems. Indicate whether the commutative, associative, distributive, or identity property applies to each equation.

① $(3)(8) = (8)(3)$ ② $(8 - 4) \times 5 = 40 - 20$

③ $12 = 1 \times 12$ ④ $5 - 3 = -3 + 5$

1.9 Order of operations

In many calculations, the answer depends on the order of operations. For example, in the expression $4 + 2 \times 3$, first multiply $2 \times 3 = 6$ and then add $4 + 6 = 10$, whereas in the similar expression $(4 + 2) \times 3$, first add $4 + 2 = 6$ and then multiply $6 \times 3 = 18$. Since one answer is 10 while the other is 18, the order of operations makes a difference.

$$4 + 2 \times 3 = 4 + 6 = 10$$
$$(4 + 2) \times 3 = 6 \times 3 = 18$$

The abbreviation **PEMDAS** shows the order of operations.

- P for parentheses. Do parentheses first.
- E for exponents. Do exponents next.
- MD for multiplication/division. Do these left to right.
- AS for addition/subtraction. Do these left to right.

Example. Evaluate each expression.

(A) $7 + 5 \times 2^2 = 7 + 5 \times 4 = 7 + 20 = 27$ (exp/mult/add)

(B) $12 \div (9 - 6) = 12 \div 3 = 4$ (par/div)

(C) $14 - \frac{(8+7)}{3} = 14 - \frac{15}{3} = 14 - 5 = 9$ (par/div/sub)

Note that $\frac{15}{3}$ means 15 divided by 3 (recall Sec. 1.7).

Problems. Evaluate each expression below.

① $(6 + 8 \div 2) \times 4 =$

② $30 - (9 - 6)^3 \div 3 =$

③ $11 - 3 \times 2 + 8 \div 2^2 =$

④ $(12 - 7)^2 - 9 \times 2 =$

⑤ $15 - \dfrac{(9+7)}{(3-1)} =$

⑥ $2 + (3 - 2^3) =$

⑦ $(9 - 4 \times 3)^3 \div 3 =$

⑧ $(4 - 3^2) - 15 \div 5 =$

⑨ $\dfrac{(6+6)}{(1+2)} - \dfrac{(15+9)}{(5-3)} =$

1.10 The distributive property

According to the distributive property:
$$a \times (b + c) = a \times b + a \times c$$
For example, $3 \times (9 - 4) = 3 \times 9 - 3 \times 4 = 27 - 12 = 15$. If you subtract first, you get the same answer: $3 \times 5 = 15$. One use of the distributive property is to multiply numbers with multiple digits. For example, consider 8×15. If we write $15 = 10 + 5$, we can use the distributive property:
$$8 \times 15 = 8 \times (10 + 5) = 8 \times 10 + 8 \times 5 = 80 + 40 = 120$$
Examples. Apply the distributive property.

(A) $6 \times 17 = 6 \times (10 + 7) = 6 \times 10 + 6 \times 7 = 60 + 42 = 102$

(B) $5 \times 29 = 5 \times (30 - 1) = 5 \times 30 - 5 \times 1 = 150 - 5 = 145$

Problems. Apply the distributive property.

① $4 \times 13 =$

② $9 \times 72 =$

③ $3 \times 99 =$

④ $45 \times 8 =$

1.11 Divisibility tests

The following tests show whether a given whole number is evenly divisible by certain whole numbers.

- An even number (ending with 0, 2, 4, 6, or 8) is evenly divisible by 2. For example, $84 \div 2 = 42$.

- If the digits add up to a multiple of 3, the number is evenly divisible by 3. For example, the digits of 84 add up to $8 + 4 = 12$, which is a multiple of 3 since $3 \times 4 = 12$. Note that $84 \div 3 = 28$.

- If the last two digits are a multiple of 4, the number is evenly divisible by 4. For example, since $4 \times 18 = 72$, 372 is divisible by 4. Note that $372 \div 4 = 93$.

- A number ending with 0 or 5 is evenly divisible by 5. For example, $60 \div 5 = 12$ and $75 \div 5 = 15$.

- A number is evenly divisible by 6 if it passes the tests for both 2 and 3. For example, 72 is even and $7 + 2 = 9$ is a multiple of 3. Note that $72 \div 6 = 12$.

- If the digits add up to a multiple of 9, the number is evenly divisible by 9. For example, the digits of 567 add up to $5 + 6 + 7 = 18$. Note that $567 \div 9 = 63$.

Problems. For each number below, is the number evenly divisible by 2, 3, 4, 5, 6, or 9? List all that apply.

① 54

② 105

③ 52

④ 96

⑤ 93

⑥ 970

⑦ 102

⑧ 198

⑨ 225

⑩ 444

⑪ 180

⑫ 386,174,952

1.12 Prime factorization

A prime number is evenly divisible only by itself and one. The first several prime numbers are:

$$2, 3, 5, 7, 11, 13, 17, 19, 23, 29, 31, 37, 41, 43, 47, 53, 59, 61, 67$$

The prime factorization of a number shows the set of prime factors that can be multiplied together to form the number. For example, the prime factorization of 42 is $2 \times 3 \times 7$ since 2, 3, and 7 are all prime numbers and $2 \times 3 \times 7 = 42$. One way to find the prime factors is to make a factor tree, like the examples below. Here is how to make a factor tree:

- Multiply any two numbers to make the given number.

- If a number is prime, circle it.

- If not, multiply two numbers to make that number.

- Repeat these steps until all of the factors are prime.

Examples. Make a factor tree for each given number.

77 = 7 × 11

190 = 19 × 2 × 5

20 = 2 × 2 × 5 = 2² × 5

Problems. Make a factor tree for each given number.

①   65   ②   18

③   24   ④   175

⑤   196   ⑥   310

Another way to find the prime factors is to make a ladder diagram. Here is how to make a ladder diagram:

- Divide the given number by a prime factor. Write the prime factor to the left of the given number and the answer below the given number.

- Continue this until the answer is also a prime factor.

Examples. Make a ladder diagram for each given number.

$$3 \,\lfloor 39 \qquad 39 \div 3 = 13$$
$$ \,\lfloor 13 \qquad 13 \text{ is prime}$$
$$39 = 3 \times 13$$

$$2 \,\lfloor 60 \qquad 60 \div 2 = 30$$
$$2 \,\lfloor 30 \qquad 30 \div 2 = 15$$
$$3 \,\lfloor 15 \qquad 15 \div 3 = 5$$
$$ \,\lfloor 5 \qquad 5 \text{ is prime}$$
$$60 = 2 \times 2 \times 3 \times 5$$
$$60 = 2^2 \times 3 \times 5$$

$$5 \,\lfloor 75 \qquad 75 \div 5 = 15$$
$$3 \,\lfloor 15 \qquad 15 \div 3 = 5$$
$$ \,\lfloor 5 \qquad 5 \text{ is prime}$$
$$75 = 3 \times 5 \times 5$$
$$75 = 3 \times 5^2$$

Problems. Make a ladder diagram for each given number.

①   63   ②   100   ③   68

1.13 Greatest common factor

The greatest common factor (GCF) of at least two numbers is the largest whole number that is a factor of each of the numbers. For example, 10 is the GCF of 30 and 40. Find the largest number that evenly divides into each given number.

Example. What is the GCF of 15 and 25?

$15 = 3 \times 5$ and $25 = 5 \times 5$. The GCF is 5.

Example. What is the GCF of 36 and 48?

$36 = 3 \times 12$ and $48 = 4 \times 12$. The GCF is 12.

Example. What is the GCF of 8, 12, and 16?

$8 = 2 \times 4$, $12 = 3 \times 4$, and $16 = 4 \times 4$. The GCF is 4.

Problems. Determine the GCF of each set of numbers.

① 12, 18   ② 27, 36

③ 22, 66   ④ 45, 75

⑤ 48, 56   ⑥ 24, 36, 42

1.14 Factoring out the GCF

Factoring is basically the distributive property backwards. Recall the distributive property (Sec. 1.10):

$$a \times (b + c) = a \times b + a \times c$$

An example of the distributive property is:

$$5 \times (2 + 7) = 5 \times 2 + 5 \times 7 = 10 + 35$$

If we write this in reverse, we call it factoring. We factor the GCF (Sec. 1.13), which is 5, out of 10 and 35 to write:

$$10 + 35 = 5 \times (2 + 7)$$

Examples. Factor out the GCF from each sum below.

(A) $16 + 20 = 4 \times 4 + 4 \times 5 = 4 \times (4 + 5)$

(B) $48 + 72 = 24 \times 2 + 24 \times 3 = 24 \times (2 + 3)$

Problems. Factor out the GCF from each sum below.

① $6 + 15 =$

② $56 + 70 =$

③ $12 + 72 =$

④ $36 + 60 + 72 =$

1.15 Least common multiple

The least common multiple (LCM) of at least two numbers is the smallest whole number that is a multiple of each of the numbers. For example, 20 is the LCM of 5 and 10. Find the smallest whole number that is evenly divisible by each given number.

Example. What is the LCM of 2 and 3?

2, 4, 6, 8 ... and 3, 6, 9, 12 ... The LCM is 6.

Example. What is the LCM of 9 and 12?

9, 18, 27, 36, 45 ... and 12, 24, 36, 48 ... The LCM is 36.

Example. What is the LCM of 8 and 10?

8, 16, 24, 32, 40 ... and 10, 20, 30, 40 ... The LCM is 40.

Problems. Determine the LCM of each set of numbers.

① 5, 6   ② 8, 12

③ 15, 20   ④ 21, 63

⑤ 16, 18   ⑥ 12, 16, 32

1.16 Square roots

Recall from Sec. 1.4 that a power of 2 is called a square. For example, three squared equals nine: $3^2 = 9$. A square root is the opposite of a square. A square root asks, "What number squared equals the number under the square root sign?" The square root sign is $\sqrt{}$. For example, $\sqrt{16} = 4$ because $4^2 = 4 \times 4 = 16$.

Examples. Find the positive square root of each number.

(A) $\sqrt{4} = 2$ because $2^2 = 2 \times 2 = 4$.

(B) $\sqrt{81} = 9$ because $9^2 = 9 \times 9 = 81$.

Problems. Find the positive square root of each number.

① $\sqrt{25} =$ ② $\sqrt{64} =$ ③ $\sqrt{49} =$

④ $\sqrt{36} =$ ⑤ $\sqrt{144} =$ ⑥ $\sqrt{1} =$

⑦ $\sqrt{0} =$ ⑧ $\sqrt{100} =$ ⑨ $\sqrt{256} =$

1.17 Factor out perfect squares

The number $\sqrt{3}$ is irrational. In contrast, $\sqrt{4}$ is rational since $\sqrt{4} = 2$ is a whole number. A number like 4, where the square root is a whole number, is called a perfect square. The number 4 is a perfect square because $2^2 = 4$. A perfect square can be factored out of some square roots, such as $\sqrt{12} = \sqrt{4 \times 3} = \sqrt{4}\sqrt{3} = 2\sqrt{3}$, where $2\sqrt{3}$ means 2 times $\sqrt{3}$.

Examples. Factor out the largest perfect square.

(A) $\sqrt{45} = \sqrt{9 \times 5} = \sqrt{9}\sqrt{5} = 3\sqrt{5}$

(B) $\sqrt{32} = \sqrt{16 \times 2} = \sqrt{16}\sqrt{2} = 4\sqrt{2}$

(C) $\sqrt{72} = \sqrt{36 \times 2} = \sqrt{36}\sqrt{2} = 6\sqrt{2}$

Problems. Factor out the largest perfect square.

① $\sqrt{18} =$

② $\sqrt{48} =$

③ $\sqrt{28} =$

④ $\sqrt{108} =$

Multiple Choice Questions

① Which of these temperatures is coldest?

   (A) 8°C   (B) 20°C   (C) −16°C   (D) 0°C   (E) −12°C

② Which of these is farthest from 0 on a number line?

   (A) 4   (B) −8   (C) 7   (D) −5   (E) −3

③ Which expression equals the greatest value?

   (A) −8 − 6   (B) −8 + 6   (C) 8 + (−6)   (D) 8 − (−6)

④ How much did a change going from $a = 10$ to $a = -15$?

   (A) −25   (B) −5   (C) 5   (D) 25   (E) 150

⑤ Given $x = 12$ and $y = -4$, what is $-x - y$?

   (A) −16   (B) −8   (C) 8   (D) 16   (E) 48

⑥ Which expression is equivalent to $(-8) \times (-8) \times (-8)$?

   (A) $(-8)^3$ (B) 8^{-3} (C) $(-3) \times (-8)$ (D) $3 \times (-8)$

⑦ Evaluate $(-5)^4$.

   (A) −625   (B) −25   (C) −20   (D) 125   (E) 625

⑧ Which expression does **NOT** equal 64?

   (A) $(-8)^2$ (B) $(-4)^3$ (C) 2^6 (D) 4^3 (E) 8^2

⑨ Order 3, $|-2|$, −1, and 1 from least to greatest.

   (A) $-1, 1, |-2|, 3$ (B) $|-2|, -1, 1, 3$ (C) $-1, |-2|, \ 1, 3$

⑩ In $50 - (4 + 6 \div 2)^2$, what must be performed first?

(A) $50 - 4$ (B) $4 + 6$ (C) $6 \div 2$ (D) 2^2

⑪ Evaluate $\frac{24-6}{6-3} - (8 - 6 \div 2)^2$.

(A) -23 (B) -19 (C) -16 (D) 1 (E) 5

⑫ Jo buys 9 eggs. Al eats 4. Ed eats 5. How many are left?

(A) $9 - (4 - 5)$ (B) $9 - (5 - 4)$ (C) $9 - (4 + 5)$ (D) $9 + (5 - 4)$

⑬ Which expression is **NOT** equivalent to $5 \times (9 - 2)$?

(A) $(9 - 2) \times 5$ (B) $45 - 10$ (C) $5 \times 9 - 5 \times 2$ (D) $5 \times 9 + 5 \times 2$

⑭ Which of these is evenly divisible by 2, 5, and 9?

(A) 1980 (B) 3645 (C) 5472 (D) 5678 (E) $10,999$

⑮ What is the prime factorization of 72?

(A) $2^2 \times 3^2$ (B) $2^2 \times 3^3$ (C) $2^3 \times 3^2$ (D) $2^3 \times 3^3$

⑯ Which of these only has two prime factors?

(A) 30 (B) 42 (C) 70 (D) 105 (E) 143

⑰ What is the greatest common factor of 48 and 80?

(A) 4 (B) 8 (C) 16 (D) 24 (E) 40

⑱ What is the least common multiple of 25 and 35?

(A) 75 (B) 175 (C) 250 (D) 350 (E) 875

⑲ Order 4^2, $\sqrt{49}$, 5^2, and 2^5 from least to greatest.

(A) $\sqrt{49}, 4^2, 5^2, 2^5$ (B) $4^2, 5^2, 2^5, \sqrt{49}$ (C) $\sqrt{49}, 2^5, 4^2, 5^2$

-2-
FRACTIONS

2.1 Reducing fractions

2.2 Mixed numbers

2.3 Lowest common denominator

2.4 Comparing fractions

2.5 Add and subtract fractions

2.6 Reciprocals

2.7 Multiply and divide fractions

2.8 Exponents of fractions

2.9 Negative exponents

2.10 Rationalize the denominator

2.11 Properties of fractions

2.12 Distributing with fractions

2.13 Factoring with fractions

2.14 Fractions over fractions

2.15 Fraction computations

2.16 Division with remainders

2.1 Reducing fractions

When the numerator and denominator of a fraction share a common factor, the fraction can be reduced to simplest form by dividing the numerator and denominator each by the greatest common factor. It may help to review Sec. 1.13. For example, $\frac{8}{12}$ can be reduced because 8 and 12 are each divisible by 4. The greatest common factor (GCF) of 8 and 12 is 4. Divide 8 and 12 each by 4 to reduce the fraction:

$$\frac{8}{12} = \frac{8 \div 4}{12 \div 4} = \frac{2}{3}$$

Examples. Reduce each fraction to its simplest form.

(A) $\frac{27}{36} = \frac{27 \div 9}{36 \div 9} = \frac{3}{4}$ 　　　　　 (B) $\frac{125}{50} = \frac{125 \div 25}{50 \div 25} = \frac{5}{2}$

Problems. Reduce each fraction to its simplest form.

① $\frac{16}{20} =$ 　　　　　　　　　② $\frac{13}{26} =$

③ $\frac{15}{10} =$ 　　　　　　　　　④ $\frac{18}{54} =$

⑤ $\frac{96}{32} =$ 　　　　　　　　　⑥ $\frac{49}{28} =$

2.2 Mixed numbers

A mixed number combines a whole number and a fraction together. For example, the mixed number $3\frac{1}{4}$ means 3 and $\frac{1}{4}$. The 3 and $\frac{1}{4}$ are added together in $3\frac{1}{4}$, such that $3\frac{1}{4}$ is larger than 3 by $\frac{1}{4}$. An alternative way to express a fraction larger than one is an improper fraction. In an improper fraction like $\frac{13}{4}$, the numerator is greater than the denominator. To convert a mixed number of the form $a\frac{b}{c}$ into an improper fraction, use the formula $a\frac{b}{c} = \frac{a \times c + b}{c}$ (see the examples).

Examples. Make equivalent improper fractions.

(A) $4\frac{3}{5} = \frac{4 \times 5 + 3}{5} = \frac{23}{5}$ $\qquad$ (B) $7\frac{1}{2} = \frac{7 \times 2 + 1}{2} = \frac{15}{2}$

Problems. Make equivalent improper fractions.

① $5\frac{2}{3} =$ $\qquad\qquad$ ② $6\frac{2}{5} =$

③ $1\frac{1}{2} =$ $\qquad\qquad$ ④ $10\frac{3}{4} =$

To convert an improper fraction of the form $\frac{a}{b}$ into a mixed number, follow these steps (illustrated in the examples):

- Perform the division $a \div b$ to get a whole number plus a remainder. For example, $39 \div 5 = 7R4$ (seven with a remainder of four) because $7 \times 5 = 35$ and $39 - 35 = 4$.

- The mixed number equals the whole number plus the remainder divided by the original denominator.

Examples. Make equivalent mixed numbers.

(A) $\frac{21}{4} = 21 \div 4 = 5R1 = 5\frac{1}{4}$ since $5 \times 4 = 20$ and $21 - 20 = 1$

(B) $\frac{31}{7} = 31 \div 7 = 4R3 = 4\frac{3}{7}$ since $4 \times 7 = 28$ and $31 - 28 = 3$

Here is an alternative way to write these examples:

(A) $\frac{21}{4} = \frac{20+1}{4} = \frac{20}{4} + \frac{1}{4} = 5\frac{1}{4}$ (B) $\frac{31}{7} = \frac{28+3}{7} = \frac{28}{7} + \frac{3}{7} = 4\frac{3}{7}$

Problems. Make equivalent mixed numbers.

① $\frac{27}{5} =$

② $\frac{10}{3} =$

③ $\frac{15}{2} =$

④ $\frac{29}{8} =$

2.3 Lowest common denominator

The lowest common denominator (LCD) is the same as the least common multiple (LCM) of two or more denominators. It may help to review Sec. 1.15. As we will explore in Sec.'s 2.4-2.5, it is useful to express two or more given fractions with their LCD when adding, subtracting, or comparing two or more fractions. Follow these steps:

- First determine the LCM (follow Sec. 1.15).
- Multiply both the numerator and denominator of each fraction by the factor needed to make the LCD.

Examples. Express each set of fractions with their LCD.

(A) Given $\frac{5}{6}$ and $\frac{4}{9}$, the LCM of 6 and 9 equals 18.

$$\frac{5}{6} = \frac{5\times3}{6\times3} = \boxed{\frac{15}{18}} \text{ and } \frac{4}{9} = \frac{4\times2}{9\times2} = \boxed{\frac{8}{18}}$$

(B) Given $\frac{3}{4}$ and $\frac{7}{3}$, the LCM of 4 and 3 equals 12.

$$\frac{3}{4} = \frac{3\times3}{4\times3} = \boxed{\frac{9}{12}} \text{ and } \frac{7}{3} = \frac{7\times4}{3\times4} = \boxed{\frac{28}{12}}$$

(C) Given $\frac{9}{10}$ and $\frac{7}{25}$, the LCM of 10 and 25 equals 50.

$$\frac{9}{10} = \frac{9\times5}{10\times5} = \boxed{\frac{45}{50}} \text{ and } \frac{7}{25} = \frac{7\times2}{25\times2} = \boxed{\frac{14}{50}}$$

Problems. Express each set of fractions with their LCD.

① $\frac{3}{8}$ and $\frac{7}{12}$

② $\frac{1}{15}$ and $\frac{2}{25}$

③ $\frac{5}{2}$ and $\frac{11}{6}$

④ $\frac{5}{24}$ and $\frac{7}{32}$

⑤ $\frac{1}{21}$ and $\frac{2}{35}$

⑥ $\frac{4}{3}$ and $\frac{8}{5}$

2.4 Comparing fractions

To compare two fractions, first express each fraction with their LCD (Sec. 2.3). Once the fractions are expressed with the same denominator, whichever fraction has the greater numerator is greater. For example, consider $\frac{5}{8}$ and $\frac{7}{12}$. The LCM of 8 and 12 equals 24. Express $\frac{5}{8}$ and $\frac{7}{12}$ with their LCD to make the equivalent fractions $\frac{5\times3}{8\times3} = \frac{15}{24}$ and $\frac{7\times2}{12\times2} = \frac{14}{24}$. Now that they share the LCD, we see that $\frac{5}{8} > \frac{7}{12}$ since $\frac{15}{24} > \frac{14}{24}$. To compare a whole number to a fraction, first divide the whole number by one. For example, $3 = \frac{3}{1}$ since $\frac{3}{1} = 3 \div 1$.

Examples. Use $>$, $<$, or $=$ to compare the given numbers.

(A) Compare $\frac{3}{14}$ and $\frac{4}{21}$. The LCM of 14 and 21 is 42.

$\frac{3}{14} = \frac{3\times3}{14\times3} = \frac{9}{42}$ and $\frac{4}{21} = \frac{4\times2}{21\times2} = \frac{8}{42}$. Since $\frac{9}{42} > \frac{8}{42}, \frac{3}{14} > \frac{4}{21}$.

(B) Compare 5 and $\frac{31}{6}$. Write $5 = \frac{5}{1}$. The LCM is 6.

$\frac{5}{1} = \frac{5\times6}{1\times6} = \frac{30}{6}$ and $\frac{31}{6} = \frac{31}{6}$. Since $\frac{30}{6} < \frac{31}{6}, 5 < \frac{31}{6}$.

(C) Compare $\frac{6}{4}$ and $\frac{9}{6}$. The LCM of 4 and 6 is 12.

$\frac{6}{4} = \frac{6\times3}{4\times3} = \frac{18}{12}$ and $\frac{9}{6} = \frac{9\times2}{6\times2} = \frac{18}{12}$. Since $\frac{18}{12} = \frac{18}{12}, \frac{6}{4} = \frac{9}{6}$.

Problems. Use $>$, $<$, or $=$ to compare the given numbers.

① $\dfrac{5}{6}$ $\dfrac{7}{9}$

② $\dfrac{7}{12}$ $\dfrac{13}{20}$

③ 6 $\dfrac{72}{12}$

④ $\dfrac{11}{36}$ $\dfrac{17}{48}$

⑤ $\dfrac{2}{11}$ $\dfrac{5}{33}$

⑥ $4\dfrac{1}{3}$ $5\dfrac{8}{9}$

2.5 Add and subtract fractions

To add or subtract two fractions, first express each fraction with their LCD (Sec. 2.3). Once the fractions have the same denominator, add or subtract the numerators. (If adding or subtracting negative values, it may help to review Sec. 1.3.) To add or subtract with a whole number and a fraction, first divide the whole number by one. For example, $8 = \frac{8}{1}$ since $\frac{8}{1} = 8 \div 1$. If a mixed number is involved, first convert the mixed number into an improper fraction (Sec. 2.2). If the answer is reducible, reduce your answer (Sec. 2.1).

Examples. Add or subtract the given numbers.

(A) $\frac{8}{15} + \frac{9}{20} = \frac{8 \times 4}{15 \times 4} + \frac{9 \times 3}{20 \times 3} = \frac{32}{60} + \frac{27}{60} = \frac{32 + 27}{60} = \frac{59}{60}$

(B) $\frac{11}{8} - \frac{4}{5} = \frac{11 \times 5}{8 \times 5} - \frac{4 \times 8}{5 \times 8} = \frac{55}{40} - \frac{32}{40} = \frac{23}{40}$

(C) $-8 + \left(-\frac{10}{3}\right) = \frac{-8}{1} + \frac{-10}{3} = \frac{-8 \times 3}{1 \times 3} + \frac{-10}{3} = \frac{-24 - 10}{3} = -\frac{34}{3}$

Notes: $8 = 8 \div 1 = \frac{8}{1}$ and $-24 + (-10) = -24 - 10$.

(D) $\frac{5}{14} - \left(-\frac{4}{21}\right) = \frac{5 \times 3}{14 \times 3} - \frac{-4 \times 2}{21 \times 2} = \frac{15}{42} - \frac{-8}{42} = \frac{15 - (-8)}{42} = \frac{23}{42}$

(E) $7\frac{5}{6} - 4\frac{2}{3} = \frac{47}{6} - \frac{14}{3} = \frac{47}{6} - \frac{14 \times 2}{3 \times 2} = \frac{47 - 28}{6} = \frac{19}{6} = 3\frac{1}{6}$

Note: $3\frac{1}{6} = \frac{3 \times 6 + 1}{6} = \frac{19}{6}$.

Problems. Add or subtract the given numbers.

① $\dfrac{3}{4} + \dfrac{5}{6} =$

② $\dfrac{7}{4} - \dfrac{1}{12} =$

③ $\dfrac{27}{5} - 3 =$

④ $\dfrac{11}{30} + \left(-\dfrac{13}{45}\right) =$

⑤ $-\dfrac{3}{14} - \dfrac{2}{35} =$

⑥ $6\dfrac{1}{3} - 3\dfrac{4}{9} =$

2.6 Reciprocals

To find the reciprocal of a fraction, swap the numerator and denominator. For example, the reciprocal of $\frac{2}{3}$ is $\frac{3}{2}$. To find the reciprocal of a whole number, divide one by the number. For example, the reciprocal of 8 is $\frac{1}{8}$ (since $8 = 8 \div 1 = \frac{8}{1}$). If the numerator of a fraction is one, its reciprocal is a whole number. For example, the reciprocal of $\frac{1}{9}$ is $\frac{9}{1} = 9 \div 1 = 9$. To find the reciprocal of a mixed number, first convert the mixed number into an improper fraction (Sec. 2.2). For example, since $3\frac{4}{5} = \frac{19}{5}$ the reciprocal of $3\frac{4}{5}$ is $\frac{5}{19}$.

Examples. Find the reciprocal of each number.

(A) $\frac{3}{7} \to \frac{7}{3}$ (B) $\frac{6}{5} \to \frac{5}{6}$ (C) $6 \to \frac{1}{6}$ (D) $\frac{1}{7} \to 7$ (E) $4\frac{1}{2} = \frac{9}{2} \to \frac{2}{9}$

Problems. Find the reciprocal of each number.

① $\frac{3}{4} \to$

② $\frac{10}{3} \to$

③ $4 \to$

④ $2\frac{2}{3} \to$

⑤ $\frac{1}{5} \to$

⑥ $7\frac{4}{5} \to$

⑦ $1 \to$

⑧ $\frac{77}{8} \to$

⑨ $3\frac{1}{3} \to$

2.7 Multiply and divide fractions

Multiplying fractions is easy: Multiply their numerators to make the new numerator, and multiply their denominators to make the new denominator. Dividing by a fraction is the same as multiplying by the reciprocal (Sec. 2.6):

$$\frac{a}{b} \div \frac{c}{d} = \frac{a}{b} \times \frac{d}{c}$$

For example, $\frac{2}{3} \div \frac{4}{5}$ is equivalent to $\frac{2}{3} \times \frac{5}{4}$ since the reciprocal of $\frac{4}{5}$ is $\frac{5}{4}$. To multiply or divide with a whole number and a fraction, first divide the whole number by one. For example, $7 = \frac{7}{1}$. If a mixed number is involved, first convert the mixed number into an improper fraction (Sec. 2.2). If the answer is reducible, reduce your answer (Sec. 2.1). If there are negative values, it may help to review Sec. 1.3.

Examples. Multiply or divide the given numbers.

(A) $\frac{2}{7} \times \frac{3}{5} = \frac{2 \times 3}{7 \times 5} = \frac{6}{35}$

(B) $5\frac{2}{3} \times 6 = \frac{17}{3} \times \frac{6}{1} = \frac{17 \times 6}{3 \times 1} = \frac{102}{3} = \frac{102 \div 3}{3 \div 3} = \frac{34}{1} = 34$

(C) $\frac{1}{2} \div \frac{3}{4} = \frac{1}{2} \times \frac{4}{3} = \frac{1 \times 4}{2 \times 3} = \frac{4}{6} = \frac{4 \div 2}{6 \div 2} = \frac{2}{3}$

(D) $\frac{5}{2} \div \left(-\frac{10}{3}\right) = \frac{5}{2} \times \left(\frac{-3}{10}\right) = \frac{5 \times (-3)}{2 \times 10} = \frac{-15}{20} = \frac{-15 \div 5}{20 \div 5} = -\frac{3}{4}$

Problems. Multiply or divide the given numbers.

① $\dfrac{7}{2} \times \dfrac{5}{6} =$

② $\dfrac{3}{8} \times \dfrac{4}{9} =$

③ $\dfrac{5}{12} \times 9 =$

④ $1\dfrac{3}{4} \times \left(-3\dfrac{2}{3}\right) =$

⑤ $\dfrac{1}{3} \div \dfrac{2}{7} =$

⑥ $\dfrac{3}{4} \div \dfrac{9}{2} =$

⑦ $4 \div \dfrac{2}{3} =$

⑧ $\left(-4\dfrac{2}{9}\right) \div 2\dfrac{2}{3} =$

⑨ $\left(-6\dfrac{1}{4}\right) \div \left(-\dfrac{3}{2}\right) =$

2.8 Exponents of fractions

It may help to review exponents in Sec. 1.4. When a fraction is raised to a whole number exponent, the exponents applies to both the numerator and denominator. For example, $\left(\frac{4}{5}\right)^3$ means $\frac{4\times4\times4}{5\times5\times5} = \frac{64}{125}$ (which is the same as $\frac{4}{5}\times\frac{4}{5}\times\frac{4}{5}$). If a mixed number has an exponent, like $\left(5\frac{3}{7}\right)^4$, first convert the mixed number into an improper fraction (Sec. 2.2). If there are negative values, it may help to review Sec. 1.3.

Examples. (A) $\left(\frac{5}{6}\right)^3 = \frac{5\times5\times5}{6\times6\times6} = \frac{125}{216}$ (B) $\left(\frac{9}{5}\right)^2 = \frac{9\times9}{5\times5} = \frac{81}{25}$

(C) $\left(3\frac{1}{3}\right)^4 = \left(\frac{10}{3}\right)^4 = \frac{10\times10\times10\times10}{3\times3\times3\times3} = \frac{10,000}{81}$

(D) $\left(-\frac{1}{7}\right)^3 = \left(\frac{-1}{7}\right)^3 = \frac{(-1)\times(-1)\times(-1)}{7\times7\times7} = -\frac{1}{343}$

Problems. Evaluate the following exponents.

① $\left(\frac{8}{7}\right)^2 =$ 　　　　　　　② $\left(\frac{3}{4}\right)^5 =$

③ $\left(-2\frac{1}{2}\right)^3 =$

④ $\left(-\frac{1}{2}\right)^8 =$

2.9 Negative exponents

If a number or fraction has a negative exponent:

$$a^{-e} = \frac{1}{a^e}$$

$$\left(\frac{a}{b}\right)^{-e} = \left(\frac{b}{a}\right)^{e} = \frac{b^e}{a^e}$$

For example, $2^{-3} = \frac{1}{2^3} = \frac{1}{2\times2\times2} = \frac{1}{8}$. Note that an exponent of

-1 is equivalent to a reciprocal. For example, $\left(\frac{5}{8}\right)^{-1} = \frac{8}{5}$. It

may help to review Sec.'s 1.3, 2.2, 2.6, and 2.8.

Examples. (A) $4^{-1} = \frac{1}{4}$ (B) $\left(\frac{1}{12}\right)^{-1} = 12$ (C) $\left(\frac{13}{27}\right)^{-1} = \frac{27}{13}$

(D) $3^{-4} = \frac{1}{3\times3\times3\times3} = \frac{1}{81}$ (E) $\left(\frac{7}{9}\right)^{-2} = \left(\frac{9}{7}\right)^{2} = \frac{9\times9}{7\times7} = \frac{81}{49}$

(F) $\left(2\frac{1}{3}\right)^{-3} = \left(\frac{7}{3}\right)^{-3} = \left(\frac{3}{7}\right)^{3} = \frac{3\times3\times3}{7\times7\times7} = \frac{27}{343}$

Problems. Evaluate the following exponents.

① $\left(\frac{5}{7}\right)^{-1} =$ ② $\left(4\frac{1}{3}\right)^{-1} =$

③ $5^{-3} =$ ④ $\left(\frac{4}{3}\right)^{-3} =$

⑤ $\left(3\frac{1}{3}\right)^{-5} =$

2.10 Rationalize the denominator

When a square root is multiplied by itself:

$$\sqrt{a}\sqrt{a} = \sqrt{a \times a} = \sqrt{a^2} = a$$

As an example, $\sqrt{9}\sqrt{9} = 3 \times 3 = 9$ since $\sqrt{9} = 3$. (It may help to review Sec. 1.16.) If a fraction includes a square root in the denominator, multiply the numerator and denominator each by the square root. This rationalizes the denominator. Note that the times symbol (×) is often not written when a whole number multiplies a square root. For example, $3\sqrt{2}$ is the same as $3 \times \sqrt{2}$.

Examples. Rationalize the denominator.

(A) $\dfrac{1}{\sqrt{3}} = \dfrac{1 \times \sqrt{3}}{\sqrt{3} \times \sqrt{3}} = \dfrac{\sqrt{3}}{3}$

(B) $\dfrac{2}{\sqrt{5}} = \dfrac{2 \times \sqrt{5}}{\sqrt{5} \times \sqrt{5}} = \dfrac{2\sqrt{5}}{5}$

Problems. Rationalize the denominator.

① $\dfrac{1}{\sqrt{2}} =$

② $\dfrac{7}{\sqrt{7}} =$

③ $\dfrac{3}{\sqrt{11}} =$

④ $\dfrac{12}{\sqrt{6}} =$

⑤ $\dfrac{50}{\sqrt{10}} =$

⑥ $\dfrac{7}{2\sqrt{14}} =$

2.11 Properties of fractions

The identity property of multiplication (Sec. 1.8), which is $a \times 1 = a$, can alternatively be expressed as:

$$a \times \left(\frac{1}{a}\right) = 1$$

This states that any nonzero number times its reciprocal equals 1. For example, $\frac{3}{4} \times \frac{4}{3} = 1$. The reciprocal can be used to express any division problem as a multiplication problem:

$$b \div a = b \times \frac{1}{a}$$

For example, $12 \div 3 = 12 \times \frac{1}{3} = 4$. When multiplying and dividing numbers, the associative property indicates that order doesn't matter. For example,

$$\frac{a \times b}{c \times d} = \frac{a}{c} \times \frac{b}{d} = \frac{a}{d} \times \frac{b}{c}$$

For example, $\frac{3 \times 8}{4 \times 9} = \frac{3}{9} \times \frac{8}{4} = \frac{1}{3} \times \frac{2}{1} = \frac{2}{3}$.

Problems. Apply the associative property.

① $\frac{5 \times 7}{7 \times 15} =$ ② $\frac{32 \times 6}{8 \times 3} =$

③ $\frac{5 \times 6}{24 \times 30} =$ ④ $\frac{7 \times 12}{3 \times 28} =$

2.12 Distributing with fractions

Recall the distributive property from Sec. 1.10:

$$a \times (b + c) = a \times b + a \times c$$

Common fractional forms of the distributive property are:

$$\frac{1}{a} \times (b + c) = \frac{b + c}{a}$$

$$\frac{a}{b} \times (c + d) = \frac{a \times c + a \times d}{b}$$

$$\frac{a}{b} \times \left(\frac{c}{d} + \frac{e}{f}\right) = \frac{a \times c}{b \times d} + \frac{a \times e}{b \times f}$$

Examples. (A) $\frac{1}{12} \times (3 + 4) = \frac{3+4}{12}$ (B) $\frac{5}{6} \times (2 + 3) = \frac{10+15}{6}$

(C) $\frac{1}{5} \times \left(\frac{1}{2} - \frac{1}{3}\right) = \frac{1}{10} - \frac{1}{15}$ (D) $\frac{6}{7} \times \left(\frac{2}{3} + \frac{4}{5}\right) = \frac{12}{21} + \frac{24}{35}$

Problems. Simplify using the distributive property. You don't need to go farther than the examples above went.

① $\frac{1}{3} \times (6 + 9) =$

② $\frac{3}{4} \times (28 - 16) =$

③ $\frac{1}{2} \times \left(\frac{1}{3} + \frac{1}{5}\right) =$

④ $\frac{3}{7} \times \left(\frac{4}{7} - \frac{2}{5}\right) =$

2.13 Factoring with fractions

Recall from Sec. 1.14 that factoring basically applies the distributive property in reverse. For example, in $\frac{1}{8} + \frac{1}{12}$, the GCF of 8 and 12 is 4, which allows us to write:

$$\frac{1}{8} + \frac{1}{12} = \frac{1}{4 \times 2} + \frac{1}{4 \times 3} = \frac{1}{4} \times \left(\frac{1}{2} + \frac{1}{3} \right)$$

Examples. Factor out the GCF.

(A) $\frac{4}{15} - \frac{3}{20} = \frac{4}{5 \times 3} - \frac{3}{5 \times 4} = \frac{1}{5} \times \left(\frac{4}{3} - \frac{3}{4} \right)$

(B) $\frac{3}{4} + \frac{9}{10} = \frac{3 \times 1}{2 \times 2} + \frac{3 \times 3}{2 \times 5} = \frac{3}{2} \times \left(\frac{1}{2} + \frac{3}{5} \right)$

(D) $\frac{9}{2} + \frac{5}{2} = \frac{9+5}{2}$

(E) $\frac{4}{7} + \frac{6}{7} = \frac{4+6}{7} = \frac{2 \times 2 + 2 \times 3}{7} = \frac{2 \times (2+3)}{7}$

Problems. Factor out the GCF like the examples above.

① $\frac{7}{30} + \frac{11}{45} =$

② $\frac{15}{16} - \frac{9}{20} =$

③ $\frac{8}{3} + \frac{12}{3} =$

④ $\frac{6}{5} + \frac{2}{5} - \frac{1}{5} =$

2.14 Fractions over fractions

Consider the fraction over the fraction below.

$$\frac{\frac{4}{5}}{\frac{2}{3}}$$

Recall from Sec. 1.7 that a fraction line represents division. Therefore, the above fraction is equivalent to $\frac{4}{5} \div \frac{2}{3}$. In Sec. 2.7, we learned that dividing by a fraction is equivalent to multiplying by its reciprocal, such that $\frac{4}{5} \div \frac{2}{3} = \frac{4}{5} \times \frac{3}{2}$.

Examples. Simplify the fractions below.

(A) $\dfrac{\frac{7}{4}}{\frac{5}{2}} = \dfrac{7}{4} \div \dfrac{5}{2} = \dfrac{7}{4} \times \dfrac{2}{5} = \dfrac{7 \times 2}{4 \times 5} = \dfrac{14}{20} = \dfrac{14 \div 2}{20 \div 2} = \dfrac{7}{10}$

(B) $\dfrac{4\frac{2}{5}}{3} = 4\dfrac{2}{5} \div 3 = \dfrac{22}{5} \div \dfrac{3}{1} = \dfrac{22}{5} \times \dfrac{1}{3} = \dfrac{22 \times 1}{5 \times 3} = \dfrac{22}{15}$

Problems. Simplify the fractions below.

① $\dfrac{\frac{9}{16}}{\frac{3}{8}} =$

② $\dfrac{8}{\frac{1}{4}} =$

③ $\dfrac{4\frac{1}{2}}{1\frac{3}{4}} =$

2.15 Fraction computations

These calculations involve a variety of methods from this chapter (and also some methods from Chapter 1).

Problems. Simplify each expression as much as possible. (In contrast, in Sec. 2.13 we left answers unsimplified.)

① $4 - \dfrac{6-3}{2+2} =$

② $6\dfrac{4}{5} - \left(-3\dfrac{2}{5}\right) =$

③ $3\dfrac{1}{6} \times \dfrac{10}{3} + \dfrac{2}{9} \div \left(\dfrac{1}{3}\right)^2 =$

④ $\left(-\dfrac{2}{3}\right)^2 - \left(-\dfrac{1}{2}\right)^3 =$

⑤ $\left(\dfrac{8}{3}\right)^{-1} + 4^{-2} =$

⑥ $\dfrac{\frac{3}{4}+\frac{2}{3}}{\frac{1}{3}-\frac{1}{4}} =$

2.16 Division with remainders

A division problem with a remainder can be expressed as a mixed number. The strategy we used in Sec. 2.2 (page 34) to convert an improper fraction into a mixed number can be used to perform division with remainders.

Examples. Express each answer as a mixed number.

(A) $32 \div 5 = 6R2 = 6\frac{2}{5}$ since $6 \times 5 = 30$ and $32 - 30 = 2$

(B) $60 \div 8 = 7R4 = 7\frac{4}{8} = 7\frac{1}{2}$ since $7 \times 8 = 56$ and $60 - 56 = 4$

and since $\frac{4}{8} = \frac{1}{2}$. (We reduced $\frac{4}{8}$ down to $\frac{1}{2}$.)

Problems. Express each answer as a mixed number.

    ① $23 \div 3 =$

    ② $68 \div 7 =$

    ③ $30 \div 4 =$

    ④ $21 \div 9 =$

    ⑤ $11 \div 2 =$

Multiple Choice Questions

① What is the lowest common denominator of $\frac{5}{6}$ and $\frac{4}{9}$?

   (A) 6   (B) 9   (C) 18   (D) 36   (E) 54

② Which of these numbers is smallest?

   (A) $\frac{3}{16}$ (B) $\frac{5}{32}$ (C) $\frac{1}{4}$ (D) $\frac{7}{8}$ (E) 1

③ Express $\frac{60}{7}$ as a mixed number.

   (A) $7\frac{5}{8}$ (B) $8\frac{1}{2}$ (C) $8\frac{4}{7}$ (D) $8\frac{6}{7}$ (E) $9\frac{3}{7}$

④ Express $4\frac{5}{6}$ as an improper fraction.

   (A) $\frac{5}{2}$ (B) $\frac{14}{3}$ (C) $\frac{17}{3}$ (D) $\frac{29}{6}$ (E) $\frac{31}{6}$

⑤ Which number is **NOT** equivalent to $\frac{21}{4}$?

   (A) $\frac{63}{12}$ (B) $5\frac{1}{4}$ (C) $\frac{105}{20}$ (D) $\frac{32}{8}$ (E) $\frac{84}{16}$

⑥ Order $\frac{17}{5}$, $\frac{11}{3}$, $3\frac{1}{2}$, and 3 from least to greatest.

(A) $3, \frac{17}{5}, \frac{11}{3}, 3\frac{1}{2}$ (B) $3, \frac{17}{5}, 3\frac{1}{2}, \frac{11}{3}$ (C) $3, 3\frac{1}{2}, \frac{17}{5}, \frac{11}{3}$ (D) $3, 3\frac{1}{2}, \frac{11}{3}, \frac{17}{5}$

⑦ What is the reciprocal of $2\frac{3}{8}$?

   (A) $\frac{4}{13}$ (B) $\frac{8}{13}$ (C) $\frac{8}{19}$ (D) $2\frac{8}{3}$ (E) $4\frac{2}{3}$

⑧ How much is 4 pies divided equally 5 ways?

   (A) $\frac{1}{4}$ pie   (B) $\frac{1}{5}$ pie   (C) $\frac{4}{5}$ pie   (D) $\frac{5}{4}$ pie   (E) $\frac{5}{9}$ pie

⑨ Evaluate $\frac{8}{3} + \frac{7}{4}$.

    (A) $\frac{13}{3}$ (B) $\frac{14}{3}$ (C) $\frac{15}{7}$ (D) $\frac{49}{12}$ (E) $\frac{53}{12}$

⑩ Evaluate $\frac{4}{5} - \frac{2}{15}$.

    (A) $\frac{2}{3}$ (B) $\frac{3}{2}$ (C) $\frac{1}{5}$ (D) $\frac{14}{15}$ (E) $\frac{8}{75}$

⑪ Evaluate $\frac{6}{7} \times \frac{4}{3}$.

    (A) $\frac{7}{5}$ (B) $\frac{8}{7}$ (C) $\frac{9}{11}$ (D) $\frac{9}{14}$ (E) $\frac{10}{21}$

⑫ Evaluate $\frac{4}{9} \div \frac{2}{3}$.

    (A) $\frac{2}{3}$ (B) $\frac{4}{9}$ (C) $\frac{10}{9}$ (D) $\frac{7}{18}$ (E) $\frac{8}{27}$

⑬ Evaluate $\left(\frac{2}{5}\right)^3$.

    (A) $\frac{6}{5}$ (B) 1 (C) $\frac{2}{15}$ (D) $\frac{4}{25}$ (E) $\frac{8}{125}$

⑭ Evaluate $\left(\frac{3}{4}\right)^{-2}$.

    (A) $-\frac{3}{2}$ (B) $\frac{1}{4}$ (C) $-\frac{3}{8}$ (D) $\frac{9}{16}$ (E) $\frac{16}{9}$

⑮ What is $\frac{3}{4}$ of 48 pens?

(A) 8 pens (B) 12 pens (C) 32 pens (D) 36 pens (E) 64 pens

⑯ Rationalize the denominator of $\frac{1}{\sqrt{6}}$.

    (A) $\sqrt{6}$ (B) $\frac{2}{\sqrt{6}}$ (C) $\frac{\sqrt{6}}{2}$ (D) $\frac{\sqrt{6}}{3}$ (E) $\frac{\sqrt{6}}{6}$

-3-

DECIMALS AND PERCENTS

3.1 Place value

3.2 Comparing decimals

3.3 Powers of ten

3.4 Rounding decimals

3.5 Add and subtract decimals

3.6 Multiply decimals

3.7 Percents

3.8 Decimals to fractions

3.9 Fractions to decimals

3.10 Repeating decimals

3.11 Divide decimals

3.12 Comparing rational numbers

3.13 Classifying numbers

3.14 Decimal computations

3.1 Place value

The example below indicates the place value of each digit of the number 1,234.567. Note that the "th" in "tenths," "hundredths," and "thousandths" is very significant.

$$1\,,2\ \ 3\ \ 4\,.5\ \ 6\ \ 7$$

thousands
hundreds
tens
units
tenths
hundredths
thousandths

Examples. What is the place value of the indicated digit?

(A) the 4 in 24.13 is in the units place

(B) the 7 in 0.79 is in the tenths place

(C) the 5 in 356.2 is in the tens place

(D) the 2 in 1.025 is in the hundredths place

Problems. What is the place value of the indicated digit?

① the 8 in 2.187

② the 6 in 625

③ the 9 in 0.94

④ the 1 in 0.00123

⑤ the 7 in 7,214.638

3.2 Comparing decimals

To compare two decimals, first check the leftmost nonzero place value of each number. If one begins at a higher place value than the other, that decimal is larger. If the leftmost nonzero place value of each number is the same, check the next digit to the right and so on. Note that trailing zeroes don't matter: $3 = 3.00$. Recall the greater than ($>$) and less than ($<$) signs.

Examples. Use $>$, $<$, or $=$ to compare the given numbers.

    (A) $0.34 > 0.339$ (B) $0.004 < 0.04$ (C) $0.2 = 0.20$

Examples. Order the numbers from least to greatest.

    (E) $0.04, 0.0008, 0.2, 0.006 \rightarrow 0.0008, 0.006, 0.04, 0.2$

    (F) $51.38, 51.34, 51.4, 51.299 \rightarrow 51.299, 51.34, 51.38, 51.4$

Problems. Use $>$, $<$, or $=$ to compare the given numbers.

    ① 0.2   0.08   ② 88.3   88.7   ③ 6.9   6.899

    ④ 4.99   5.42   ⑤ 7.0   7.00   ⑥ 0.009   0.01

Order the given numbers from least to greatest.

    ⑦ $0.3, 0.12, 0.23, 0.194$

    ⑧ $2.703, 2.698, 2.710, 2.6$

3.3 Powers of ten

A positive exponent (or power) means to multiply the base times itself that many times (Sec. 1.4). If the base is 10, a positive power indicates how many zeroes will follow the 1. For example, $10^4 = 10 \times 10 \times 10 \times 10 = 10,000$ has 4 zeroes. If the power is negative, use the formula from Sec. 2.9:

$$10^{-p} = \frac{1}{10^p}$$

For example, $10^{-4} = \frac{1}{10^4} = \frac{1}{10,000} = 0.0001$. Note that there are $4 - 1 = 3$ zeroes between the decimal point and the 1 in the case of a negative power. Note that $10^0 = 1$ (Sec. 1.4).

Examples. (A) $10^1 = 10$ (B) $10^4 = 10,000$ (C) $10^{-2} = 0.01$

Problems. Evaluate the following powers of ten.

① $10^3 =$ ② $10^{-3} =$ ③ $10^2 =$

④ $10^6 =$ ⑤ $10^0 =$ ⑥ $10^{-1} =$

⑦ $10^{-5} =$ ⑧ $10^8 =$ ⑨ $10^{-4} =$

⑩ $10^7 =$ ⑪ $10^{-6} =$ ⑫ $10^9 =$

3.4 Rounding decimals

To round a decimal, look at the digit that is one place to the right of the rounded digit. For example, when rounding to the nearest tenth, look at the hundredths place, and when rounding to the nearest hundredth, look at the thousandths place. Recall from Sec. 1.5 that 1 thru 4 round down while 5 thru 9 round up.

Examples. Round each number to the nearest tenth.

(A) 0.18 rounds to 0.2 because 8 is 5 or more.

(B) 0.12 rounds to 0.1 because 2 is less than 5.

(C) 0.15 rounds to 0.2 because 5 is 5 or more.

Examples. Round each number to the nearest hundredth.

(A) 6.743 rounds to 6.74 because 3 is less than 5.

(B) 6.747 rounds to 6.75 because 7 is 5 or more.

Problems. Round each number to the nearest tenth.

① 0.349 ≈   ② 3.792101 ≈   ③ 0.55 ≈

Problems. Round each number to the nearest hundredth.

④ 0.016 ≈   ⑤ 7.0646 ≈   ⑥ 3.725 ≈

3.5 Add and subtract decimals

If one number has fewer decimal places than the other, add trailing zeroes to that number until both numbers have the same number of digits after the decimal point. When you add or subtract the decimals, preserve the placement of the decimal point, like the examples below.

Examples. Add or subtract the decimals.

(A) $4.236 + 3.45$ $\rightarrow$

$$\begin{array}{r} 4.236 \\ +\ 3.450 \\ \hline 7.686 \end{array}$$

(B) $6 - 0.72$ $\rightarrow$

$$\begin{array}{r} 6.00 \\ -\ 0.72 \\ \hline 5.28 \end{array}$$

(C) $0.17 + 0.08$ $\rightarrow$

$$\begin{array}{r} 0.17 \\ +\ 0.08 \\ \hline 0.25 \end{array}$$

(D) $1.89 - 1.7$ $\rightarrow$

$$\begin{array}{r} 1.89 \\ -\ 1.70 \\ \hline 0.19 \end{array}$$

Problems. Add or subtract the decimals.

① $2.3 + 3.8 =$

② $4.5 - 1.75 =$

③ $0.7 + 0.04 =$

④ $0.06 - 0.005 =$

⑤ $9.86 + 0.7 =$

⑥ $1 - 0.003 =$

3.6 Multiply decimals

Follow these steps to multiply two decimals:

- How many decimal places does each number have? Add these together.

- Ignore the decimal places and multiply.

- Shift the decimal point such that the answer has the same number of decimal places as the first step.

Examples. Multiply the decimals.

$$
\text{(A)} \quad \begin{array}{r} 2.4 \\ \times\ 1.5 \\ \hline \end{array} \quad \xrightarrow{\ 2\ \text{places}\ } \quad \begin{array}{r} 24 \\ \times\ 15 \\ \hline 120 \\ 240 \\ \hline 360 \end{array} \qquad \begin{array}{r} 2.4 \\ \times\ 1.5 \\ \hline 3.60 \end{array} \qquad \begin{array}{r} 2.4 \\ \times\ 1.5 \\ \hline 3.6 \end{array}
$$

Note that $3.60 = 3.6$ (trailing zeroes don't matter).

The final answer is 3.6.

$$
\text{(B)} \quad \begin{array}{r} 1.47 \\ \times\ 0.2 \\ \hline \end{array} \quad \xrightarrow{\ 3\ \text{places}\ } \quad \begin{array}{r} 147 \\ \times\ 2 \\ \hline 294 \end{array} \qquad \begin{array}{r} 1.47 \\ \times\ 0.2 \\ \hline 0.294 \end{array}
$$

The final answer is 0.294.

$$
\text{(C)} \quad \begin{array}{r} 0.124 \\ \times\ 0.3 \\ \hline \end{array} \quad \xrightarrow{\ 4\ \text{places}\ } \quad \begin{array}{r} 124 \\ \times\ 3 \\ \hline 372 \end{array} \qquad \begin{array}{r} 0.124 \\ \times\ 0.3 \\ \hline 0.0372 \end{array}
$$

The final answer is 0.0372.

Problems. Multiply the decimals.

①
$$\begin{array}{r} 7.3 \\ \times\ 0.6 \\ \hline \end{array}$$

②
$$\begin{array}{r} 4.3 \\ \times\ 1.2 \\ \hline \end{array}$$

③
$$\begin{array}{r} 0.382 \\ \times\ \ 0.4 \\ \hline \end{array}$$

④
$$\begin{array}{r} 32 \\ \times\ 0.3 \\ \hline \end{array}$$

⑤
$$\begin{array}{r} 0.93 \\ \times\ 0.8 \\ \hline \end{array}$$

⑥
$$\begin{array}{r} 0.27 \\ \times\ 0.19 \\ \hline \end{array}$$

⑦
$$\begin{array}{r} 8.4 \\ \times\ 0.21 \\ \hline \end{array}$$

⑧
$$\begin{array}{r} 0.56 \\ \times\ \ 7 \\ \hline \end{array}$$

⑨
$$\begin{array}{r} 0.015 \\ \times\ 0.6 \\ \hline \end{array}$$

⑩
$$\begin{array}{r} 0.008 \\ \times\ 0.03 \\ \hline \end{array}$$

3.7 Percents

A percent is a fraction of one hundred. For example, 32% means 32 out of 100. Note that 100% = 1 because 100 out of 100 equals $\frac{100}{100} = 1$. When there is a specific value, we use the term "percent," but for an unspecified amount we use the term "percentage." Compare "five percent of the class" to "a percentage of the class." Converting between decimals and percents is easy:

- To convert a decimal to a percent, multiply by 100%.
- To convert a percent to a decimal, divide by 100%.

Examples. (A) $0.47 \times 100\% = 47\%$ (B) $\frac{125\%}{100\%} = 1.25$

Problems. Convert each number to a percent.

① 0.74 =   ② 0.08 =   ③ 1.9 =

④ 0.3 =   ⑤ 4 =   ⑥ 0.006 =

Problems. Convert each percent to a decimal.

⑦ 81% =   ⑧ 230% =   ⑨ 9% =

⑩ 70% =   ⑪ 0.2% =   ⑫ 5.6% =

3.8 Decimals to fractions

To convert a decimal to a fraction, remove the decimal point and divide by the power of ten corresponding to the final decimal position of the given decimal. If possible, reduce the answer according to Sec. 2.1.

Examples. Convert each decimal to a fraction.

(A) $0.8 = \frac{8}{10} = \frac{8 \div 2}{10 \div 2} = \frac{4}{5}$ 　　　　　 (B) $0.57 = \frac{57}{100}$

(C) $0.675 = \frac{675}{1000} = \frac{675 \div 25}{1000 \div 25} = \frac{27}{40}$ 　 (D) $3.3 = \frac{33}{10}$

Problems. Convert each decimal to a fraction.

① $0.49 =$ 　　　　　　　　② $0.4 =$

③ $0.005 =$ 　　　　　　　④ $1.5 =$

⑤ $0.75 =$ 　　　　　　　⑥ $0.024 =$

⑦ $2.88 =$ 　　　　　　　⑧ $0.96 =$

3.9 Fractions to decimals

To convert a fraction to a decimal, multiply the numerator and denominator by the factor needed to make an equivalent fraction with a denominator equal to 10, 100, 1000, etc. The power indicates the final decimal position. For example, $\frac{731}{100}$ ends in the hundredths place: $\frac{731}{100} = 7.31$.

Examples. Convert each fraction to a decimal.

(A) $\frac{3}{5} = \frac{3 \times 2}{5 \times 2} = \frac{6}{10} = 0.6$ 　　 (B) $\frac{7}{20} = \frac{7 \times 5}{20 \times 5} = \frac{35}{100} = 0.35$

(C) $\frac{7}{4} = \frac{7 \times 25}{4 \times 25} = \frac{175}{100} = 1.75$ 　　 (D) $\frac{1}{8} = \frac{1 \times 125}{8 \times 125} = \frac{125}{1000} = 0.125$

Problems. Convert each fraction to a decimal.

① $\frac{3}{4} =$ 　　　　　　　　② $\frac{27}{50} =$

③ $\frac{7}{2} =$ 　　　　　　　　④ $\frac{16}{125} =$

⑤ $\frac{31}{25} =$ 　　　　　　　　⑥ $\frac{11}{5} =$

⑦ $\frac{17}{40} =$ 　　　　　　　　⑧ $\frac{543}{200} =$

3.10 Repeating decimals

A bar over a digit indicates that the digit repeats forever. For example, $0.\overline{4} = 0.444444444444...$ A bar over a group of digits indicates that the group repeats forever. For example, $0.\overline{28} = 0.282828282828...$ To convert a repeating decimal to a fraction, find a suitable example below.

Examples. Convert each decimal to a fraction.

(A) $0.\overline{3} = \frac{3}{9} = \frac{3 \div 3}{9 \div 3} = \frac{1}{3}$ 　　　　(B) $0.\overline{41} = \frac{41}{99}$

(C) $0.\overline{321} = \frac{321}{999} = \frac{321 \div 3}{999 \div 3} = \frac{107}{333}$ 　(D) $5.\overline{2} = 5 + 0.\overline{2} = 5\frac{2}{9} = \frac{47}{9}$

(E) $0.0\overline{7} = \frac{1}{10}(0.\overline{7}) = \frac{1}{10}\left(\frac{7}{9}\right) = \frac{7}{90}$

Problems. Convert each decimal to a fraction.

① $0.\overline{24} =$ 　　　　　　　② $0.\overline{6} =$

③ $0.\overline{05} =$ 　　　　　　　④ $0.0\overline{5} =$

⑤ $0.\overline{110} =$ 　　　　　　　⑥ $1.\overline{3} =$

If the denominator can't be factored exclusively in terms of 2's and 5's (because $2 \times 5 = 10$ and the decimal system is based on ten), the fraction converts to a repeating decimal. To convert a fraction to a repeating decimal, find a suitable example below.

Examples. Convert each fraction to a decimal.

(A) $\frac{1}{3} = \frac{1 \times 3}{3 \times 3} = \frac{3}{9} = 0.\overline{3}$

(B) $\frac{6}{11} = \frac{6 \times 9}{11 \times 9} = \frac{54}{99} = 0.\overline{54}$

(C) $\frac{82}{333} = \frac{82 \times 3}{333 \times 3} = \frac{246}{999} = 0.\overline{246}$

(C) $\frac{8}{90} = \frac{1}{10}\left(\frac{8}{9}\right) = \frac{0.\overline{8}}{10} = 0.0\overline{8}$

(D) $\frac{1}{300} = \frac{1 \times 3}{300 \times 3} = \frac{3}{900} = \frac{1}{100}\left(\frac{3}{9}\right) = \frac{1}{100}\left(0.\overline{3}\right) = 0.00\overline{3}$

(E) $\frac{100}{99} = 1\frac{1}{99} = 1 + \frac{1}{99} = 1 + 0.\overline{01} = 1.\overline{01}$

(F) $\frac{5}{3} = 1\frac{2}{3} = 1 + \frac{2}{3} = 1 + \frac{2 \times 3}{3 \times 3} = 1 + \frac{6}{9} = 1 + 0.\overline{6} = 1.\overline{6}$

Problems. Convert each fraction to a decimal.

① $\frac{25}{33} =$

② $\frac{14}{111} =$

③ $\frac{1}{30} =$

④ $\frac{14}{9} =$

⑤ $\frac{12}{11} =$

⑥ $\frac{1}{6} =$

3.11 Divide decimals

One way to divide two decimals is to express the division as a fraction. For example, $0.3 \div 0.25 = \frac{0.3}{0.25}$. Next multiply the numerator and denominator by a power of ten sufficient to remove the decimal points. For example, $\frac{0.3}{0.25} = \frac{0.3 \times 100}{0.25 \times 100} = \frac{30}{25}$. Then convert the fraction to a decimal (Sec.'s **3.9** and **3.10**).

Examples. Divide the decimals.

(A) $0.2 \div 0.8 = \frac{0.2 \times 10}{0.8 \times 10} = \frac{2}{8} = \frac{2 \times 125}{8 \times 125} = \frac{250}{1000} = 0.250 = 0.25$

(B) $4.5 \div 0.15 = \frac{4.5 \times 100}{0.15 \times 100} = \frac{450}{15} = \frac{450 \div 15}{15 \div 15} = \frac{30}{1} = 30$

(C) $0.345 \div 0.5 = \frac{0.345 \times 1000}{0.5 \times 1000} = \frac{345}{500} = \frac{345 \times 2}{500 \times 2} = \frac{690}{1000} = 0.690 = 0.69$

Problems. Divide the decimals.

① $0.3 \div 2.5 =$

② $0.32 \div 0.4 =$

③ $1 \div 0.05 =$

④ $2.48 \div 0.8 =$

3.12 Comparing rational numbers

To compare fractions, decimals, or percents, convert each number to the same form (such as decimal). It may help to review Sec.'s 3.7 thru 3.10.

Examples. Use >, <, or = to compare the given numbers.

(A) $\frac{1}{5} > 0.18$ since $\frac{1}{5} = \frac{1 \times 2}{5 \times 2} = \frac{2}{10} = 0.2 > 0.18$

(B) $70\% < 0.72$ since $70\% = \frac{70\%}{100\%} = 0.7 < 0.72$

(C) $\frac{5}{4} = 125\%$ since $\frac{5}{4} = \frac{5 \times 25}{4 \times 25} = \frac{125}{100} = 1.25 = 125\%$

Problems. Use >, <, or = to compare the given numbers.

① $1.4 \qquad \frac{3}{2}$

② $0.5\% \qquad 0.02$

③ $\frac{7}{20} \qquad 30\%$

Order the given numbers from least to greatest.

④ $\frac{3}{5}, 0.53, 35\%$

⑤ $\frac{5}{2}, 2.7, 120\%$

3.13 Classifying numbers

Whole numbers include 0, 1, 2, 3, 4, 5, 6, and so on. Some books and instructors distinguish between whole numbers and natural numbers, in which whole numbers include the 0 but natural numbers don't include the 0. Some books and instructors also distinguish these between integers, with the difference that integers include negative numbers. Not all books and instructors agree on these definitions, so it helps to pay attention in class.

Rational numbers include fractions and decimals, like $\frac{2}{3}$ and 1.24, in addition to integers. Repeating decimals, like $0.\overline{3} = 0.333333333333333...$, are included with rational numbers. Irrational numbers go on forever without repeating, like $\sqrt{2} = 1.41421356...$ and $\pi = 3.14159265...$

Problems. Is each number whole, rational, or irrational?

① 27.2   ② 4,315,986   ③ $\frac{1}{3}$

④ $\sqrt{3}$ ⑤ $\frac{12}{4}$ ⑥ $\sqrt{16}$

3.14 Decimal computations

These calculations involve a variety of methods from this chapter (and also some methods from Chapters 1 and 2).

Problems. Evaluate each expression.

① $3.4 + (-0.72) - (-1.45) =$

② $12 \times 0.3 - \dfrac{1}{2} \times 10 \times 0.3^2 =$

③ $8.1 - (1.75 + 0.2 \div 0.8)^2 =$

④ $(0.4)^{-1} + (0.2)^{-1} =$

⑤ $1.6 \times \dfrac{7 - 3.8}{0.65 + 0.15} =$

⑥ $10^{-6} \times \dfrac{10^4 \times 10^5}{(10^3)^2} =$

Multiple Choice Questions

① What is the decimal place of the 7 in 283.179?

(A) units   (B) tenths   (C) hundredths   (D) thousandths

② Order 0.6, 0.079, 0.008, 0.097, and 0.005 from least to greatest.

(A) 0.005, 0.6, 0.079, 0.008, 0.097 (B) 0.005, 0.008, 0.079, 0.097, 0.6
(C) 0.005, 0.079, 0.008, 0.097, 0.6 (D) 0.005, 0.008, 0.6, 0.079, 0.097

③ Evaluate 10^{-3}.

(A) 0.0001   (B) 0.001   (C) −30   (D) 300   (E) 1000

④ Round 0.045736 to the nearest thousandth.

(A) 0.04   (B) 0.05   (C) 0.045   (D) 0.046   (E) 0.0457

⑤ Evaluate $0.34 + 0.016$.

(A) 0.3416   (B) 0.356   (C) 0.4   (D) 0.5   (E) 0.56

⑥ Evaluate $3.1 - 0.24$.

(A) 0.7   (B) 1.3   (C) 2.84   (D) 2.86   (E) 2.96

⑦ Evaluate 7.4×0.03.

(A) 0.0222   (B) 0.222   (C) 2.22   (D) 22.2   (E) 222

⑧ Evaluate $0.48 \div 0.2$.

(A) 0.12   (B) 0.24   (C) 0.42   (D) 1.2   (E) 2.4

⑨ Convert 0.375 to a percent.

(A) 0.00375% (B) 0.0375% (C) 3.75% (D) 37.5% (E) 375%

⑩ Convert 140% to a decimal.

(A) 0.14　(B) 1.4　(C) 14　(D) 1,400　(E) 14,000

⑪ Convert $\frac{3}{5}$ to a decimal.

(A) 0.06　(B) $0.1\overline{3}$　(C) 0.3　(D) 0.6　(E) $1.\overline{3}$

⑫ Convert 1.36 to a fraction.

(A) $\frac{7}{5}$　(B) $\frac{6}{25}$　(C) $\frac{9}{25}$　(D) $\frac{34}{25}$　(E) $\frac{63}{50}$

⑬ Which number is **NOT** equivalent to $\frac{59}{50}$?

(A) 1.18　(B) 11.8%　(C) $1\frac{9}{50}$　(D) $\frac{118}{100}$　(E) 118%

⑭ Which number is irrational?

(A) $0.2\overline{3}$　(B) 12.5%　(C) $\frac{11}{200}$　(D) 5,432　(E) $\sqrt{10}$

⑮ Order $\frac{27}{8}$, 3, 3.38, $3\frac{1}{2}$, and 337% from least to greatest.

(A) $3, 337\%, \frac{27}{8}, 3.38, 3\frac{1}{2}$　(B) $3, \frac{27}{8}, 337\%, 3.38, 3\frac{1}{2}$

(C) $3, 337\%, \frac{27}{8}, 3\frac{1}{2}, 3.38$　(D) $\frac{27}{8}, 3, 3.38, 3\frac{1}{2}, 337\%$

⑯ What is 32% of $\frac{4}{5}$?

(A) 0.128　(B) 0.256　(C) 0.32　(D) 12.8　(E) 25.6

⑰ A number that can be expressed as $\frac{x}{y}$ is _____?

(A) a whole number (B) an integer (C) a natural number

(D) a rational number (E) an irrational number

⑱ What is 6.4 MB divided equally among 16 people?

(A) 0.4 MB (B) 2.4 MB (C) 2.5 MB (D) 4 MB (E) 25 MB

⑲ If three out of four students at a school take the bus, what percent of the students **DON'T** take the bus?

(A) 4%   (B) 12%   (C) 25%   (D) 33%   (E) 75%

⑳ If a student answers 39 out of 50 questions correctly on a test, what percent of the student's answers are correct?

(A) 11%   (B) 22%   (C) 39%   (D) 61%   (E) 78%

-4-

PROPORTIONS

4.1 Ratios and fractions

4.2 Reducing ratios

4.3 Ratio tables

4.4 Applying ratios

4.5 The part and the whole

4.6 Rates and fractions

4.7 Rate tables

4.8 Rates and decimals

4.9 Constant speed

4.10 The rate pyramid

4.11 Applying the pyramid

4.12 Applying rates

4.13 Linear proportions

4.14 Applying proportions

4.1 Ratios and fractions

A ratio expresses a fixed relationship as a fraction in the form $a{:}b$. The ratio $a{:}b$ is equivalent to the fraction $\frac{a}{b}$. The first number is the numerator and the second number is the denominator.

Examples. Express each ratio as a fraction.

(A) $2{:}9 = \frac{2}{9}$ (B) $5{:}3 = \frac{5}{3}$ (C) $12{:}1 = \frac{12}{1} = 12$

Problems. Express each ratio as a fraction.

① $2{:}3 =$ ② $1{:}5 =$ ③ $6{:}11 =$

④ $4{:}1 =$ ⑤ $3{:}8 =$ ⑥ $14{:}5 =$

Examples. Express each fraction as a ratio.

(A) $\frac{1}{6} = 1{:}6$ (B) $\frac{5}{11} = 5{:}11$ (C) $\frac{7}{4} = 7{:}4$

Problems. Express each fraction as a ratio.

⑦ $\frac{3}{8} =$ ⑧ $\frac{1}{2} =$ ⑨ $\frac{9}{4} =$

⑩ $\frac{1}{4} =$ ⑪ $\frac{3}{2} =$ ⑫ $\frac{11}{16} =$

4.2 Reducing ratios

When the numerator and denominator of a ratio share a common factor, the ratio can be reduced to simplest form by dividing the numerator and denominator each by the greatest common factor. It may help to review Sec.'s 1.13 and 2.1. For example, 9:6 can be reduced because 9 and 6 are each divisible by 3. The greatest common factor (GCF) of 9 and 6 is 3. Divide 9 and 6 each by 3 to reduce the ratio:

$$9{:}6 = \frac{9}{6} = \frac{9 \div 3}{6 \div 3} = \frac{3}{2} = 3{:}2$$

Example. $28{:}8 = \frac{28}{8} = \frac{28 \div 4}{8 \div 4} = \frac{7}{2} = 7{:}2$

Problems. Reduce each ratio to its simplest form.

① 4:12 =

② 15:10 =

③ 18:24 =

④ 35:21 =

⑤ 30:6 =

4.3 Ratio tables

A ratio table includes equivalent ratios. Put the numerator on the top row and the denominator on the bottom row. Put the original ratio at the left. Multiply (or divide) both the numerator and denominator of the original ratio by the same number to make equivalent ratios.

cats	3	6	9	12	15
dogs	4	8	12	16	20

Example. In the table above, the original ratio is 3:4. The other columns were made by multiplying the numerator and denominator of the original ratio by 2, 3, 4, and 5. This shows that 3:4, 6:8, 9:12, 12:16, and 15:20 are all equivalent.

Problem. ① Complete a table for 5 apples to 2 oranges.

apples	5				
oranges	2				

Problem. ② Fill in the missing elements of the table below.

wins		18			45
losses			15		25

4.4 Applying ratios

An equivalent ratio can be found by multiplying or dividing the numerator and denominator by the same number. For example, 2:3 is equivalent to 8:12 because:

$$2{:}3 = \frac{2}{3} = \frac{2 \times 4}{3 \times 4} = \frac{8}{12} = 8{:}12$$

Example. A girl uses 8 strawberries and 25 blueberries to make one smoothie. If she makes several smoothies using 56 strawberries, how many blueberries does she need?

$$\frac{56 \text{ strawberries}}{8 \text{ strawberries}} = 56 \div 8 = 7 \ (\textbf{NOT} \text{ the final answer})$$

Multiply 8 strawberries by 7 to make 56 strawberries.

$$8{:}25 = \frac{8}{25} = \frac{8 \times 7}{25 \times 7} = \frac{56}{175} = 56{:}175 \rightarrow \boxed{175 \text{ blueberries}}$$

Example. A company manufactured 8000 t.v.'s and 35 are defective. If a store purchases 1600 of these t.v.'s, about how many of the purchased t.v.'s will be defective?

$$\frac{8000 \text{ t.v.'s}}{1600 \text{ t.v.'s}} = 8000 \div 1600 = 5 \ (\textbf{NOT} \text{ the final answer})$$

Divide 8000 t.v.'s by 5 to make 1600 t.v.'s.

$$8000{:}35 = \frac{8000}{35} = \frac{8000 \div 5}{35 \div 5} = \frac{1600}{7} = 1600{:}7 \rightarrow \boxed{7 \text{ t.v.'s}}$$

Problems. ① The ratio of girls to boys at a school is 4:3. If there are 60 girls at the school, how many boys are there?

② A school needs 12 bus drivers to transport 960 students. How many bus drivers are needed to transport 240 students?

③ There are 320 cows and 48 horses on a ranch. Express the ratio of cows to horses in reduced form.

4.5 The part and the whole

A ratio can be expressed as part to part, part to whole, or whole to part. For example, suppose a box contains 9 blue crayons and 7 red crayons, such that there are $9 + 7 = 16$ crayons in total:

- The ratio of blue crayons to red crayons is 9:7.
- The ratio of blue crayons to all of the crayons is 9:16.
- The ratio of all of the crayons to blue crayons is 16:9.

The first ratio above is part to part, the second is part to whole, and the third is whole to part.

Example. There are 56 trucks and 112 cars in a parking lot. What is the ratio of cars to automobiles?

56 trucks + 112 cars = 168 automobiles (**NOT** the final answer)

$$\frac{\text{cars}}{\text{automobiles}} = \frac{112}{168} = \frac{112 \div 56}{168 \div 56} = \frac{2}{3} = \boxed{2:3}$$

Example. The ratio of total apples to red apples in a barrel is 5:3. There are 80 apples in the barrel. How many are red?

$$\frac{80 \text{ total apples}}{5 \text{ total apples}} = 80 \div 5 = 16 \text{ (\textbf{NOT} the final answer)}$$

$$\frac{\text{total apples}}{\text{red apples}} = \frac{5}{3} = \frac{5 \times 16}{3 \times 16} = \frac{80}{48} = 80:48 \rightarrow \boxed{48 \text{ red apples}}$$

Problems. ① A family gathering includes 70 children and 28 adults. Express the ratio of children to family members in reduced form.

② The ratio of pens to writing utensils in a box is 4:7. There are 140 writing utensils in the box. How many are pens?

③ The ratio of desktops to computers in a warehouse is 3:4. The warehouse has 120 desktops. How many computers are in the warehouse?

4.6 Rates and fractions

A rate is a fraction made by dividing quantities that have different units. Examples of rates include $\frac{200 \text{ miles}}{3 \text{ hours}}$, $\frac{50 \text{ dollars}}{7 \text{ days}}$, and $\frac{77 \text{ sales}}{8 \text{ hours}}$. It is common to express a rate as a decimal, such as $3.2 \frac{\text{meters}}{\text{second}}$. Sometimes, a rate works out to a whole number, like $40 \frac{\text{boxes}}{\text{truck}}$. (For a whole number or decimal, the denominator isn't plural. It's 40 boxes per 1 truck since $40 = \frac{40}{1}$.)

Examples. Express each rate as a fraction. Include units.

(A) 300 kilometers in 7 hours $= \frac{300 \text{ kilometers}}{7 \text{ hours}}$

(B) 5 dollars for 2 yards $= \frac{5 \text{ dollars}}{2 \text{ yards}}$

Problems. Express each rate as a fraction. Include units.

① 800 miles in 9 hours =

② 20 minutes for 3 problems =

③ 24 seeds in 1 package =

④ 5 dollars for 12 eggs =

4.7 Rate tables

A rate table includes equivalent rates. Put the numerator on the top row and the denominator on the bottom row. Put the original rate at the left. Multiply (or divide) both the numerator and denominator of the original rate by the same number to make equivalent rates.

miles	50	100	150	200	250
gallons	3	6	9	12	15

Example. In the table above, the original rate is $\frac{50 \text{ miles}}{3 \text{ gallons}}$. The other columns were made by multiplying the numerator and denominator of the original rate by 2, 3, 4, and 5. This shows that $\frac{50}{3}$, $\frac{100}{6}$, $\frac{150}{9}$, $\frac{200}{12}$, and $\frac{250}{15}$ are all equivalent.

Problem. ① Complete a table for 125 dollars for 8 hours.

dollars	125				
hours	8				

Problem. ② Fill in the missing elements of the table below.

beats	5			20	25
seconds			12		20

4.8 Rates and decimals

A rate can be expressed as a decimal, just as any fraction can be converted to a decimal. It may help to review Sec.'s 3.9 and 3.10.

Examples. Convert each rate to a decimal.

(A) $\dfrac{24 \text{ grams}}{5 \text{ servings}} = \dfrac{24 \times 2 \text{ grams}}{5 \times 2 \text{ servings}} = \dfrac{48 \text{ grams}}{10 \text{ servings}} = 4.8$ grams per serving

(B) $\dfrac{908 \text{ words}}{25 \text{ minutes}} = \dfrac{908 \times 4 \text{ words}}{25 \times 4 \text{ minutes}} = \dfrac{3632 \text{ words}}{100 \text{ minutes}} = 36.32$ words per minute

(C) $\dfrac{7 \text{ dollars}}{2 \text{ tickets}} = \dfrac{7 \times 5 \text{ dollars}}{2 \times 5 \text{ tickets}} = \dfrac{35 \text{ dollars}}{10 \text{ tickets}} = 3.5$ dollars per ticket

Problems. Convert each rate to a decimal.

① 3 inches in 50 days =

② 210 miles in 4 hours =

③ 94 dollars for 10 hours =

④ 7 loaves in 20 minutes =

4.9 Constant speed

If an object travels with constant speed, the distance (d), time (t), and speed (which is a rate, r) are related by $r = \frac{d}{t}$. This formula can be written three different ways, depending on which quantity you are solving for:

$$r = \frac{d}{t} \quad , \quad d = r \times t \quad , \quad t = \frac{d}{r}$$

Example. A car travels 60 mph for 7 hours. How far does the car travel?

$$d = r \times t = 60 \frac{\text{miles}}{\text{hour}} \times 7 \text{ hours} = \boxed{420 \text{ miles}}$$

Example. A boy travels 800 meters with a speed of 4 m/s. How much time does the trip take?

$$t = \frac{d}{r} = \frac{800 \text{ meters}}{4 \text{ meters/second}} = \boxed{200 \text{ seconds}}$$

Note that the meters cancel and that $\frac{1}{1/\text{sec}} = 1 \div \frac{1}{\text{sec}} = 1 \times \frac{\text{sec}}{1} = 1$ sec. (To divide by a fraction, multiply by its reciprocal.)

Example. A bug crawls 30 inches in 6 minutes. What is the bug's speed?

$$r = \frac{d}{t} = \frac{30 \text{ inches}}{6 \text{ minutes}} = \boxed{5 \text{ inches per minute}}$$

Problems. ① A train travels 280 miles in 7 hours. What is the speed of the train?

② A bicycle travels with a speed of 18 km/hr for 1.5 hours. How far does the bicycle travel?

③ A bird flies 200 yards with a speed of 8 yards per second. How much time does this take?

4.10 The rate pyramid

Recall the forms of the equation for constant speed:

$$r = \frac{d}{t} \quad , \quad d = r \times t \quad , \quad t = \frac{d}{r}$$

All three forms are illustrated by the rate pyramid below.

Cover the quantity that you're solving for.

- When you cover d, since r and t are beside one another they are multiplied: $d = r \times t$.

- When you cover r, since d is over t you get $r = \frac{d}{t}$.

- When you cover t, since d is over r you get $t = \frac{d}{r}$.

Example. Given $P = \frac{W}{t}$, solve for W.

Since W is over t, make a pyramid with W at the top.

When you cover W, since P is beside t you get $W = P \times t$.

Example. Given $P = \frac{W}{t}$, solve for t.

When you cover t, since W is over P you get $t = \frac{W}{P}$.

Problems. ① Given $C = \frac{Q}{V}$, use a pyramid to solve for Q.

② Given $V = I \times R$, use a pyramid to solve for I.

③ Given $n = \frac{c}{v}$, use a pyramid to solve for v.

4.11 Applying the pyramid

If the rate equation is given, you can use the rate pyramid to solve for an unknown.

Example. Given that $d = \frac{m}{V}$, where d is density, m is mass, and V is volume, determine the mass of an object that has a volume of 5 cc and a density of 18 g/cc.

Since m is over V, make a pyramid with m at the top.

When you cover m, since d is beside V you get $m = d \times V$.

$$m = d \times V = 18\frac{\text{g}}{\text{cc}} \times 5 \text{ cc} = \boxed{90 \text{ g}}$$

Example. Given that $I = P \times r$, where I is interest, P is the principal, and r is rate, determine the rate if the principal is \$200 and the interest earned is \$8.

Since P multiplies r, make a pyramid with P beside r.

When you cover r, since I is over P you get $r = \frac{I}{P}$.

$$r = \frac{I}{P} = \frac{\$8}{\$200} = \frac{8 \div 2}{200 \div 2} = \frac{4}{100} = 0.04 = \boxed{4\%}$$

Problems. ① Given that $M = \frac{I}{O}$, where M is magnification, I is image distance, and O is object distance, determine the image distance if the object distance is 40 cm and $M = 8$.

② Given that $m = \frac{rise}{run}$, where m is the slope, determine the run if the rise is -12 and the slope is 4.

4.12 Applying rates

An equivalent rate can be found by:

- If necessary, convert the rate to a fraction (Sec. 3.8).

- Multiply or divide the numerator and denominator by the same number.

For example, 0.6 m/s is equivalent to $\frac{12}{20}$ m/s because:

$$0.6 \text{ m/s} = \frac{6 \text{ m}}{10 \text{ s}} = \frac{6 \text{ m} \div 2}{10 \text{ s} \div 2} = \frac{3 \text{ m}}{5 \text{ s}} = \frac{3 \text{ m} \times 4}{5 \text{ s} \times 4} = \frac{12 \text{ m}}{20 \text{ s}}$$

Example. A student reads 6.4 pages per minute. How many pages can the student read in 15 minutes?

$$6.4 \frac{\text{pgs.}}{\text{min.}} = \frac{64 \text{ pgs.}}{10 \text{ min.}} = \frac{64 \text{ pgs.} \div 2}{10 \text{ min.} \div 2} = \frac{32 \text{ pgs.}}{5 \text{ min.}}$$

$$= \frac{32 \text{ pgs.} \times 3}{5 \text{ min.} \times 3} = \frac{96 \text{ pgs.}}{15 \text{ min.}}$$

In 15 minutes, the student can read $\boxed{96 \text{ pages}}$.

Example. A woman earns \$900 per week. How much does she earn in 7 weeks?

$$\$900 \text{ /wk.} = \frac{\$900}{1 \text{ wk.}} = \frac{\$900 \times 7}{1 \text{ wk.} \times 7} = \frac{\$6300}{7 \text{ wk.}}$$

In 7 weeks, the woman earns $\boxed{\$6300}$.

Problems. ① A secretary types 80 words per minute. How many words can the secretary type in 7 minutes?

② A wheel rotates 3.25 times per second. How many times does the wheel rotate in 24 seconds?

③ A car gets 42.5 miles per gallon of fuel. How far can the car travel on 8 gallons of fuel?

4.13 Linear proportions

A proportion expresses an equality between two ratios or two rates (for quantities that obey a linear proportion). For example, suppose that a loaf of bread has 16 slices and that a package of cheese has 12 slices. The ratio of slices of bread to slices of cheese is 16:12, which reduces to 4:3. We can set up a proportion stating that the ratio 16:12 equals 4:3.

$$\frac{16}{12} = \frac{4}{3}$$

Note that $\frac{16}{12} = \frac{16 \div 4}{12 \div 4} = \frac{4}{3}$.

Examples. Determine the unknown in each proportion.

(A) $\frac{2}{3} = \frac{\square}{15}$ Since $\frac{2}{3} = \frac{2 \times 5}{3 \times 5} = \frac{10}{15}$, it follows that $\square = 10$.

(B) $\frac{7}{4} = \frac{56}{\square}$ Since $\frac{7}{4} = \frac{7 \times 8}{4 \times 8} = \frac{56}{32}$, it follows that $\square = 32$.

(C) $\frac{\square}{10} = \frac{4}{8}$ We will do this in two steps.

First, reduce $\frac{4}{8} = \frac{4 \div 4}{8 \div 4} = \frac{1}{2}$. Next, $\frac{1}{2} = \frac{1 \times 5}{2 \times 5} = \frac{5}{10}$.

Now compare $\frac{5}{10} = \frac{4}{8}$ to see that $\square = 5$.

Problems. Determine the unknown in each proportion.

① $\dfrac{4}{3} = \dfrac{\square}{24}$

② $\dfrac{\square}{8} = \dfrac{20}{32}$

③ $\dfrac{7}{11} = \dfrac{21}{\square}$

④ $\dfrac{11}{\square} = \dfrac{220}{100}$

⑤ $\dfrac{\square}{36} = \dfrac{75}{45}$

4.14 Applying proportions

When quantities obey a linear proportion, you may set up a proportion in order to solve for an unknown. Be careful that both numerators (and both denominators) correspond to the same units. For example, if you travel 9 miles in 2 hours and wish to figure out how far you would travel in 6 hours at the same rate, be sure to put 2 hours and 6 hours both in a denominator. (The answer is 9 mi.× 3 = 27 mi.)

$$\frac{9 \text{ miles}}{2 \text{ hours}} = \frac{\square}{6 \text{ hours}}$$

Example. A man pays \$8 for 3 t-shirts. How much would the man pay for 12 t-shirts?

$$\frac{\$8}{3 \text{ t-shirts}} = \frac{\square}{12 \text{ t-shirts}}$$

Since $\frac{\$8}{3 \text{ t-shirts}} = \frac{\$8 \times 4}{3 \text{ t-shirts} \times 4} = \frac{\$32}{12 \text{ t-shirts}}$, the man would pay $\boxed{\$32}$.

Example. A woman plays 18 holes of golf in 3 hours. About how long would it take her to play 6 holes?

$$\frac{18 \text{ holes}}{3 \text{ hours}} = \frac{6 \text{ holes}}{\square}$$

Since $\frac{18 \text{ holes}}{3 \text{ hours}} = \frac{18 \text{ holes} \div 3}{3 \text{ hours} \div 3} = \frac{6 \text{ holes}}{1 \text{ hour}}$, it would take about $\boxed{1 \text{ hour}}$.

Problems. ① If 3 jars hold 100 beads, how many beads can 18 jars hold?

② For every 150 days of work, a woman earns 7 vacation days. How much does she need to work to earn 21 vacation days?

③ An editor charges $429 to proofread 200 pages. What would it cost to have 800 pages proofread?

Multiple Choice Questions

① Which ratio is the same as 24 apples to 18 bananas?

(A) 3:4   (B) 4:3   (C) 6:3   (D) 6:4   (E) 8:9

② Which ratio is **NOT** equivalent to 9:15?

(A) 2:3   (B) 3:5   (C) 6:10   (D) 12:20   (E) 21:35

③ The ratio of dresses to shirts in a closet is 3:2. There are 18 dresses in the closet. How many shirts are there?

(A) 6   (B) 9   (C) 12   (D) 27   (E) 36

④ A juice recipe requires 5 apples to make 2 cups of juice. How many apples are needed to make 16 cups of juice?

(A) 3   (B) 8   (C) 10   (D) 40   (E) 80

⑤ A pet clinic has 40 dogs and 25 cats. What is the ratio of cats to the total number of pets?

(A) 5:8   (B) 5:13   (C) 8:13   (D) 8:5   (E) 13:5

⑥ A student answered 7 out of 10 questions correctly on a test with 50 questions. How many answers were correct?

(A) 35   (B) 40   (C) 42   (D) 45   (E) 70

⑦ Which rate is the same as 8 inches of snow in 12 months?

(A) $\dfrac{2 \text{ in.}}{3 \text{ mo.}}$ (B) $\dfrac{3 \text{ in.}}{2 \text{ mo.}}$ (C) $\dfrac{3 \text{ in.}}{4 \text{ mo.}}$ (D) $\dfrac{4 \text{ in.}}{3 \text{ mo.}}$ (E) $\dfrac{2 \text{ in.}}{5 \text{ mo.}}$

⑧ Which rate is the same as 108 words per 25 seconds?

(A) 1.08   (B) 2.16   (C) 2.7   (D) 4.32   (E) 5.4

⑨ A car travels 60 mph (miles per hour) for 8 hours. How far does the car travel?

(A) 0.13 mi. (B) 7.5 mi. (C) 240 mi. (D) 420 mi. (E) 480 mi.

⑩ A girl travels a distance of 240 m with speed of 3 m/s. How much time does this take?

(A) 0.0125 s   (B) 8 s   (C) 72 s   (D) 80 s   (E) 720 s

⑪ A bug crawls 32 inches in 4 hours. What is the speed?

(A) 0.125 in./hr (B) 8 in./hr (C) 9 in./hr (D) 128 in./hr

⑫ A service charges $8.25 per month. How much will 6 months cost?

(A) $4.85   (B) $4.95   (C) $48.50   (D) $49.50   (E) $485

⑬ An artist can draw 7 sketches in 30 minutes. How many sketches can the artist draw in 90 minutes?

(A) 4   (B) 21   (C) 63   (D) 210   (E) 630

⑭ What does ☐ equal in the equation $\frac{3}{4} = \frac{☐}{48}$?

(A) 12   (B) 16   (C) 36   (D) 64   (E) 144

⑮ What does ☐ equal in the equation $\frac{6}{☐} = \frac{24}{72}$?

(A) 4   (B) 12   (C) 18   (D) 36   (E) 144

⑯ A map is scaled such that 3 inches represents 12 miles. What does 4.5 inches represent?

(A) 4 mi. (B) 13.5 mi. (C) 18 mi. (D) 36 mi. (E) 54 mi.

⑰ A sign says, "buy 3, get 2 free." How many would you need to buy in order to get 10 free?

(A) 5   (B) 15   (C) 20   (D) 30   (E) 60

⑱ A poet writes 27 lines in 36 minutes. At this rate, how many lines can the poet write in 60 minutes?

(A) 24   (B) 45   (C) 48   (D) 80   (E) 90

⑲ It takes 5 hours to deliver 125 newspapers. How long would it take to deliver 175 newspapers?

(A) 3 hr   (B) 5.5 hr   (C) 6 hr   (D) 7 hr   (E) 35 hr

⑳ A car traveling with constant speed travels 180 miles in 4 hours. How far would it travel in 5 hours?

(A) 144 mi. (B) 200 mi. (C) 215 mi. (D) 220 mi. (E) 225 mi.

-5-

VARIABLES

5.1 Variables and constants

5.2 Operations with variables

5.3 Expressions and equations

5.4 Words and variables

5.5 Using formulas

5.6 The simplest equations

5.7 Combine like terms

5.8 Isolate the unknown

5.9 Fractions and signs

5.10 Variables with exponents

5.11 Distributing with variables

5.12 Factoring with variables

5.13 Square roots

5.14 Variable in a denominator

5.15 Cross multiplying

5.16 Special solutions

5.17 Inequalities

5.1 Variables and constants

In math, a letter like x or y represents a variable. The value of the variable is generally different from one problem to another. For example, $x = 4$ satisfies the equation $3x = 12$ whereas $x = 7$ satisfies the equation $x + 9 = 16$. A variable is an unknown quantity. A numerical value like 3 or 7.2 is called a constant. A constant that multiplies a variable is called a coefficient. Since x is commonly used as a variable, to avoid confusion we don't use $\times$ for multiplication when working with variables. For example, $6x$ means 6 times x and $3(x - 4)$ means 3 times $(x - 4)$.

Example. In the equation $4y - 7 = 1$, y is the variable, 4 is the coefficient, and 4, 7, and 1 are constants.

Problems. Identify the variable, constant, or coefficient.

① What is the variable in $4(t + 2) = 20$?

② What are the constants in $3x + 2 = 8$?

③ What is the coefficient in $-5x + 22 = 7$?

④ What are the variables in $4y - 6z = 11$?

⑤ What are the coefficients in $8x + 9 = y$?

5.2 Operations with variables

Quantities that are next to one another are multiplying. For example, $3xy$ means 3 times x times y and $(x - 2)(4x)$ means $(x - 2)$ times $(4x)$. The fraction line represents division. For example, $\frac{5x}{2}$ means $5x$ divided by 2.

Examples. Identify the arithmetic operations in words.

(A) $8x - 7$ means "8 times x minus 7."

(B) $5(x + 3)$ means "5 times quantity x plus 3."

(C) $(x + 1)(x - 1)$ means "quantity x plus 1 times quantity x minus 1."

(D) $\frac{x+2}{4}$ means "quantity x plus 2 divided by 4."

Problems. Identify the arithmetic operations in words.

① $3x^2 + 2x - 1$

② $-x + \frac{9}{x}$

③ $x(x + 6)$

④ $\frac{x^2-4}{2x+3}$

⑤ $(2x + 1)^2(3x)$

5.3 Expressions and equations

An expression doesn't include an equal sign (=) or inequality (> or <). An expression can be simplified (but not solved). For example, the expression $5x + 3x - 2 - 4$ simplifies to $8x - 6$. Following are examples of expressions.

- $3x + 2$
- $(x^2 - 1)^3$
- $\frac{2x+5}{4}$

An equation includes one equal sign (=). An equation can be solved. For example, the equation $2x - 1 = 5$ is solved by $x = 3$ because $2(3) - 1 = 6 - 1 = 5$. Following are examples of equations.

- $5x - 4 = 6$
- $(x - 2)(x + 3) = 0$
- $\frac{x-2}{3} = 4$

Problems. Is it an expression or an equation?

① $x - 1 = 1$

② $(x + 3)(2x + 1)$

③ $x^2 - 3x + 2$

④ $(3x - 2)^2 = 100$

⑤ $\frac{x}{2} = \frac{7}{6}$

⑥ $\frac{4x-3}{x} - \frac{x}{2}$

5.4 Words and variables

Expressions with variables are used to represent ideas.

Examples. Write an expression to represent each idea.

(A) three more than a number: $x + 3$

(B) two years younger than Julie's age: $x - 2$

(C) five times as many cookies: $5x$

(D) one-half base times height: $\frac{1}{2}xy$ (or $\frac{1}{2}bh$)

(E) the sum of two numbers: $x + y$

(F) ten percent of the students: $0.1x$ (since $10\% = 0.1$)

Problems. Write an expression to represent each idea.

① five less than a number

② twice as many coins

③ four times as old as Mark

④ the square root of a number

⑤ distance over time squared

⑥ the product of two numbers

5.5 Using formulas

Many calculations involve plugging values into a formula. For example, the formula for the perimeter of a rectangle is $P = 2L + 2W$. If the length and width are $L = 3$ and $W = 4$ meters, the perimeter is $P = 2(3) + 2(4) = 6 + 8 = 14$ meters.

Examples. Plug the values into the given formula.

(A) $A = \frac{1}{2}bh$, $b = 6$, $h = 8$: $A = \frac{1}{2}(6)(8) = \frac{48}{2} = 24$

(B) $P = I^2R$, $I = 3$, $R = 7$: $P = (3)^2(7) = (9)(7) = 63$

Problems. Plug the values into the given formula.

① $S = 6L^2$, $L = 3$

② $v = \frac{d}{t}$, $d = 42$, $t = 7$

③ $P = b^c$, $b = 10$, $c = 4$

④ $a = \frac{1}{2}gt^2$, $g = 10$, $t = 4$

⑤ $I = \frac{V}{R}$, $V = 12$, $R = 6$

⑥ $w = \frac{xy}{z}$, $x = 2$, $y = 6$, $z = 4$

5.6 The simplest equations

Let's begin with equations that are simple enough to solve in a single step. In these equations, do the opposite of what is being done to the variable to both sides of the equation. As examples, in $x + 8 = 15$ subtract 8 from both sides, and in $3x = 6$ divide both sides by 3. (Subtraction is the opposite of addition and division is the opposite of multiplication.)

Examples. Solve for the variable in each equation.

(A) In $x + 8 = 15$, subtract 8 from both sides.

We get $x + 8 - 8 = 15 - 8$, which becomes $x = \boxed{7}$.

Check the answer: $x + 8 = 7 + 8 = 15$.

(B) In $3x = 6$, divide both sides by 3.

We get $\frac{3x}{3} = \frac{6}{3}$, which becomes $x = \boxed{2}$.

Check the answer: $3x = 3(2) = 6$.

(C) In $x - 4 = 2$, add 4 to both sides.

We get $x - 4 + 4 = 2 + 4$, which becomes $x = \boxed{6}$.

Check the answer: $x - 4 = 6 - 4 = 2$.

(D) In $\frac{x}{5} = 3$, multiply both sides by 5.

We get $\frac{x}{5} 5 = 3(5)$, which becomes $x = \boxed{15}$.

Check the answer: $\frac{x}{5} = \frac{15}{5} = 3$.

Problems. Solve for the variable in each equation.

① $x - 6 = 3$

② $2x = 10$

③ $\dfrac{x}{4} = 9$

④ $x + 2 = 5$

⑤ $x + 9 = 14$

⑥ $x - 4 = 8$

⑦ $7x = 49$

⑧ $\dfrac{x}{8} = 7$

5.7 Combine like terms

The equation $3x - 2 + 4x + 7 = 9x - 3$ has six terms, which are separated by $-$, $+$, and $=$ signs. The terms $3x$, $4x$, and $9x$ are like terms because they have the same power of the variable, and 2, 7, and 3 are like terms because they are all constant terms. In the expression $4x^2 - 2x + 5x^2 - 6x$, the terms $4x^2$ and $5x^2$ are like terms because they both have x^2, and $2x$ and $6x$ are like terms. Like terms can be combined by factoring out the variable: $4x^2 + 5x^2 = (4 + 5)x^2 = 9x^2$.

Examples. (A) $3x - 2 + 4x + 7 = (3 + 4)x - 2 + 7 = 7x + 5$

(B) $9x^2 + 8x - 4x^2 + x = (9 - 4)x^2 + (8 + 1)x = 5x^2 + 9x$

Problems. Simplify each expression.

① $8x + 3 - 5x + 6$　　　　② $2x^2 + 4x + x^2 + 3x$

③ $9x^2 - 9 - 7x^2 - 7$　　　　④ $x^2 + 2x + 3 + 4x^2 - 2x - 1$

5.8 Isolate the unknown

In many equations, it is possible to isolate the unknown by combining like terms. First bring all of the variable terms to one side of the equation and all of the constant terms to the other side. Do this by adding or subtracting the same expression on both sides of the equation. For example, in $9x - 4 = 3x + 8$, subtract $3x$ from both sides and add 4 to both sides. Note that we do the opposite of what is already being done: subtract $3x$ because it is being added to 8, but add 4 because it is being subtracted from $9x$. After isolating the unknown, divide by its coefficient. For example, in $6x = 12$, divide both sides by 6 to get $x = \frac{12}{6} = 2$.

Examples. Solve for the variable in each equation.

(A) $5x + 3 = 9 + 2x$

Subtract 3 and $2x$.

$5x - 2x = 9 - 3$

$3x = 6$

Divide by 3.

$x = \frac{6}{3} = \boxed{2}$

Check: $5(2) + 3 = 13 = 9 + 2(2)$.

(B) $52 - 3x = 4x + 10$

Add $3x$ and subtract 10.

$52 - 10 = 4x + 3x$

$42 = 7x$

Divide by 7.

$\frac{42}{7} = \boxed{6} = x$

Check: $52 - 3(6) = 34 = 4(6) + 10$.

Problems. Solve for the variable in each equation.

① $9x + 1 = 7x + 7$

② $3x - 8 = 2x - 4$

③ $5x - 9 = 3x + 9$

④ $x + 3 = 2x - 5$

⑤ $8x - 24 = 7x - 3x$

⑥ $2x - 6 = 12$

⑦ $6x - 3 + 4x = 7x + 9$

⑧ $50 - 8x = 15 - 3x$

5.9 Fractions and signs

In some equations, the answer turns out to be a fraction, the constants are fractions, or the answer is negative. It may help to review Sec.'s 2.5 and 2.7.

Examples. Solve for the variable in each equation.

(A) $7x + 2 = 10 + 4x$

Subtract 2 and $4x$.

$7x - 4x = 10 - 2$

$3x = 8$

Divide by 3.

$x = \boxed{\dfrac{8}{3}}$

Check: $7\left(\frac{8}{3}\right) + 2 = \frac{62}{3} = 10 + 4\left(\frac{8}{3}\right)$.

(B) $4x - \dfrac{1}{4} = 2x + \dfrac{3}{2}$

Add $\dfrac{1}{4}$ and subtract $2x$.

$4x - 2x = \dfrac{3}{2} + \dfrac{1}{4}$

$2x = \dfrac{3}{2}\left(\dfrac{2}{2}\right) + \dfrac{1}{4} = \dfrac{6}{4} + \dfrac{1}{4} = \dfrac{7}{4}$

Divide by 2.

$x = \dfrac{7}{4} \div 2 = \dfrac{7}{4} \div \dfrac{2}{1} = \dfrac{7}{4} \times \dfrac{1}{2} = \boxed{\dfrac{7}{8}}$

Check: $4\left(\frac{7}{8}\right) - \frac{1}{4} = \frac{13}{4} = 2\left(\frac{7}{8}\right) + \frac{3}{2}$.

(C) $3x + 2 = 5x + 8$

Subtract $3x$ and 8.

$2 - 8 = 5x - 3x$

$-6 = 2x$

Divide by 2.

$-\dfrac{6}{2} = \boxed{-3} = x$

Check: $3(-3) + 2 = -7 = 5(-3) + 8$.

(D) $-x - 1 = x + 1$

Add x and subtract 1.

$-1 - 1 = x + x$

$-2 = 2x$

Divide by 2.

$-\dfrac{2}{2} = \boxed{-1} = x$

Check: $-(-1) - 1 = 0 = -1 + 1$.

Problems. Solve for the variable in each equation.

① $8x + 2 = 6x + 9$

② $x - \frac{1}{3} = \frac{1}{6}$

③ $2x + 8 = 0$

④ $3x - 4 = 7x + 6$

⑤ $3x + \frac{1}{2} = \frac{1}{4} + 5x$

⑥ $5 - 6x = 7 - 3x$

⑦ $\frac{x}{2} + 8 = \frac{x}{3} + 6$

⑧ $\frac{3x}{4} + \frac{1}{2} = \frac{2x}{3} - \frac{5}{6}$

5.10 Variables with exponents

The following rules apply when working with exponents.

$$x^m x^n = x^{m+n} \quad , \quad x^0 = 1 \text{ (if } x \neq 0) \quad , \quad x^1 = x$$

$$x^{-m} = \frac{1}{x^m} \quad , \quad \frac{x^m}{x^n} = x^{m-n} \quad , \quad (ax)^m = a^m x^m$$

$$(x^m)^n = x^{mn} \quad , \quad x^{1/2} = \sqrt{x} \quad , \quad x^{m/n} = \left(\sqrt[n]{x}\right)^m$$

Examples. Simplify each expression.

(A) $x^3 x^2 = x^{3+2} = x^5$ (B) $\dfrac{x^6}{x^4} = x^{6-4} = x^2$

(C) $(x^6)^2 = x^{6(2)} = x^{12}$ (D) $x^4 x^{-4} = x^{4+(-4)} = x^0 = 1$

(E) $(4x)^3 = 4^3 x^3 = 64 x^3$ (F) $9^{3/2} = \left(\sqrt{9}\right)^3 = 3^3 = 27$

Problems. Simplify each expression.

① $x^5 x^3 =$ ② $\dfrac{x^8}{x^7} =$

③ $(x^2 x)^4 =$ ④ $(2x^3)^5 =$

⑤ $4^{3/2} =$ ⑥ $64^{2/3} =$

⑦ $\dfrac{16x^8 x^6}{(2x^5)^2} =$ ⑧ $\dfrac{x^{12} x^{-8}}{x^{-9}} =$

5.11 Distributing with variables

Recall the distributive property (Sec. 1.10):

$$x(y + z) = xy + xz$$

Examples: Apply the distributive property.

(A) $x^2(x^3 + x^2) = x^2(x^3) + x^2(x^2) = x^5 + x^4$

(B) $4(x + 3) = 4(x) + 4(3) = 4x + 12$

(C) $3x^2(x - 2) = 3x^2(x) - 3x^2(2) = 3x^3 - 6x^2$

(D) $-2(x^2 - x) = -2(x^2) - 2(-x) = -2x^2 + 2x$

Problems. Apply the distributive property.

① $4(3x - 4) =$

② $x(x + 5) =$

③ $8x^4(3x^2 - 4x) =$

④ $-3(2x - 6) =$

⑤ $5x^3(4x^6 - 3x^4 + x^2) =$

An important application of the distributive property is the f.o.i.l. method (f.o.i.l. stands for first outside inside last).

$$(w + x)(y + z) = wy + wz + xy + xz$$

Examples. Apply the f.o.i.l. method.

(A) $(x + 2)(x + 3) = x^2 + 3x + 2x + 6 = x^2 + 5x + 6$

(B) $(x - 4)^2 = (x - 4)(x - 4) = x^2 - 4x - 4x + 16 = x^2 - 8x + 16$

Problems. Apply the f.o.i.l. method.

① $(x + 7)(x + 6) =$

② $(x + 5)(x - 5) =$

③ $(x + 5)^2 =$

④ $(4x + 3)(2x - 6) =$

⑤ $(x^2 - 4x)(x + 5) =$

⑥ $(3x + 2y)(4x - y) =$

5.12 Factoring with variables

Factoring is basically the distributive property backwards. Recall the distributive property (Sec. 5.11):

$$x(y + z) = xy + xz$$

Examples. Factor out the GCF.

(A) $6x^8 + 9x^6 = 3x^6(2x^2 + 3)$

(B) $15x - 20 = 5(3x - 4)$

(C) $-x^6 - x^5 = -x^5(x + 1)$

Problems. Factor out the GCF.

① $12x^2 - 18 =$

② $8x^3 + 12x^2 =$

③ $27x^4 - 36x =$

④ $-x^{12} - x^{10} =$

⑤ $9x^7 - 6x^5 + 3x^3 =$

5.13 Square roots

The following rules apply when working with square roots.

$$\left(\sqrt{x}\right)^2 = \sqrt{x}\sqrt{x} = x \quad , \quad \sqrt{ax} = \sqrt{a}\sqrt{x}$$

Examples. (A) $\sqrt{x}\left(\sqrt{x} + 1\right) = \sqrt{x}\sqrt{x} + \sqrt{x} = x + \sqrt{x}$

(B) $\left(\sqrt{x} + 1\right)\left(\sqrt{x} - 2\right) = \sqrt{x}\sqrt{x} - 2\sqrt{x} + \sqrt{x} - 2$ (factor)

$= \sqrt{x}\sqrt{x} + (-2 + 1)\sqrt{x} + \sqrt{x} - 2 = x - \sqrt{x} - 2$

(C) $x\sqrt{x} - 4\sqrt{x} = \sqrt{x}(x - 4)$

Problems. Apply the distributive property.

① $4\sqrt{x}\left(2x - 3\sqrt{x}\right) =$

② $\sqrt{3x}\left(\sqrt{x} - \sqrt{3}\right) =$

③ $\left(\sqrt{x} + 3\right)\left(2 - \sqrt{x}\right) =$

Problems. Factor out the GCF.

④ $12x - 8\sqrt{x} =$

⑤ $x^2\sqrt{x} + x\sqrt{x} =$

5.14 Variable in a denominator

If there is a variable in a denominator, isolate the variable term and then take the reciprocal of both sides. To add or subtract fractions, first find a common denominator.

Examples. Solve for the variable in each equation.

(A) $\dfrac{1}{6} + \dfrac{1}{x} = \dfrac{1}{2}$

$\dfrac{1}{x} = \dfrac{1}{2} - \dfrac{1}{6} = \dfrac{3}{6} - \dfrac{1}{6} = \dfrac{2}{6} = \dfrac{1}{3}$

$x = \boxed{3}$

Check: $\dfrac{1}{6} + \dfrac{1}{3} = \dfrac{1}{6} + \dfrac{2}{6} = \dfrac{3}{6} = \dfrac{1}{2}$.

(B) $\dfrac{3}{2x} - \dfrac{1}{4} = \dfrac{1}{x}$

$\dfrac{3}{2x} - \dfrac{1}{x} = \dfrac{1}{4}$

$\dfrac{3}{2x} - \dfrac{2}{2x} = \dfrac{1}{2x} = \dfrac{1}{4}$

$2x = 4$

$x = \dfrac{4}{2} = \boxed{2}$

Check: $\dfrac{3}{2(2)} - \dfrac{1}{4} = \dfrac{3}{4} - \dfrac{1}{4} = \dfrac{2}{4} = \dfrac{1}{2}$.

Problems. Solve for the variable in each equation.

① $\dfrac{1}{x} - \dfrac{2}{3} = \dfrac{1}{12}$

② $\dfrac{4}{3x} + \dfrac{1}{36} = \dfrac{3}{2x}$

5.15 Cross multiplying

Equations that have the following structure can be solved by cross multiplying: Multiply along the diagonals.

$$\frac{w}{x} = \frac{y}{z} \quad \rightarrow \quad \frac{w}{x} \diagup \frac{y}{z} \quad \rightarrow \quad wz = xy$$

Examples. Solve for the variable in each equation.

(A) $\frac{3}{4} = \frac{15}{x}$

$3x = 4(15)$

$3x = 60$

$x = \frac{60}{3} = \boxed{20}$

Check: $\frac{15}{x} = \frac{15}{20} = \frac{3}{4}$

(B) $\frac{6}{x} = \frac{24}{20}$

$6(20) = 24x$

$120 = 24x$

$\frac{120}{24} = \boxed{5} = x$

Check: $\frac{6}{5} = \frac{6}{5}\left(\frac{4}{4}\right) = \frac{24}{20}$

Problems. Solve for the variable in each equation.

① $\frac{9}{4} = \frac{x}{8}$

② $\frac{8}{7} = \frac{24}{x}$

③ $\frac{18}{x} = \frac{15}{20}$

④ $\frac{x}{9} = \frac{8}{6}$

5.16 Special solutions

Some equations have no solution, multiple answers, or all real numbers solve the equation.

- $x + 4 = x + 9$ has no solution. If you subtract x from both sides, the variable cancels out and you get $4 = 9$, but 4 doesn't equal 9.

- $x^2 - 3x = 0$ has two answers: $x = 0$ and $x = 3$. You can find both answers by factoring: $x(x - 3) = 0$. Since x times $(x - 3)$ equals zero, either $x = 0$ or $x - 3 = 0$.

- All real numbers solve $x + x = 2x$ because it simplifies to $2x = 2x$, which is true for any value of x.

Problems. Find all of the solutions to each equation.

① $5x - 2x = 3x$

② $6 - 4x = 8 - 4x$

③ $x^2 = 7x$

④ $x + 3 = 5 + x - 2$

⑤ $8x + 1 - 6x = 2x + 5$

⑥ $\dfrac{3x}{2} = \dfrac{x^2}{4}$

5.17 Inequalities

An inequality like $2x + 3 < 7$ can be solved by isolating the unknown with one important exception: If you multiply or divide by a negative number, you must reverse the direction of the inequality. For example, $-x < 4$ becomes $x > -4$. For example, suppose that $x = 6$. Note that $6 > -4$ and $-6 < 4$.

Examples. Isolate the variable in each inequality.

(A) $2x + 3 < 7$

$2x < 7 - 3$

$2x < 4$

$x < \dfrac{4}{2}$

$x < 2$

(B) $3 - 2x > 9$

$-2x > 9 - 3$

$-2x > 6$

$\dfrac{-2x}{-2} < \dfrac{6}{-2}$

The $>$ sign becomes $<$ when we divide by negative two.

$x < -3$

Note: Since $x < -3$, the answer to problem (B) is negative. Therefore, $-2x$ is positive and $3 - 2x$ is also positive. This is how $3 - 2x$ is greater than 9 even though x is less than -3. There is another way to see this. In $-2x > 6$, we could add $2x$ to get $0 > 6 + 2x$ and subtract 6 to get $-6 > 2x$, which becomes $-3 > x$. Note that this is consistent with $x < -3$.

Problems. Isolate the variable in each inequality.

① $4x - 5 > 7$

② $-x < 9$

③ $6 + 3x < 5x$

④ $3 - 3x > 9$

⑤ $-4x > 9 + 5x$

⑥ $8 - 3x < 2 - 4x$

⑦ $\frac{x}{4} < -5$

⑧ $35 > -\frac{5x}{2}$

Multiple Choice Questions

① What is the coefficient in $8 + 3x = 26$?

(A) 8   (B) 3   (C) x (D) $3x$ (E) 26

② Which of the following is **NOT** an equation?

(A) $x^2 + 4x - 3$ (B) $3x = 6$ (C) $x - 1 = 1$ (D) $7 = y$ (E) $\frac{2}{x} = \frac{6}{9}$

③ Which represents seven less than a number?

(A) $\frac{x}{7}$ (B) $7x$ (C) $x + 7$ (D) $7 - x$ (E) $x - 7$

④ Which represents y cookies divided evenly 5 ways?

(A) $\frac{y}{5}$ (B) $5y$ (C) $y + 5$ (D) $5 - y$ (E) $y - 5$

⑤ Which represents the difference between two numbers?

(A) $\frac{x}{y}$ (B) $\frac{y}{x}$ (C) xy (D) $x + y$ (E) $x - y$

⑥ Given $t = 3$, evaluate $a = 2t^2 - 5t + 5$.

(A) $a = -4$ (B) $a = -1$ (C) $a = 2$ (D) $a = 3$ (E) $a = 8$

⑦ Given $Q = 24$ and $V = 3$, evaluate $C = \frac{Q}{V}$.

(A) $C = 0.125$ (B) $C = 7.2$ (C) $C = 8$ (D) $C = 62$ (E) $C = 72$

⑧ Which is equivalent to $9x - 3 - 5x + 1 + x - 8$?

(A) $4x - 11$ (B) $5x - 12$ (C) $4x - 12$

(D) $5x - 10$ (E) $-6x$

⑨ Solve for x in the equation $2 - 4x = 9$.

(A) $x = -\frac{4}{11}$ (B) $x = -\frac{7}{4}$ (C) $x = -\frac{11}{4}$ (D) $x = \frac{7}{4}$ (E) $x = \frac{11}{4}$

⑩ Solve for y in the equation $5y - 12 = 36 - y$.

(A) $y = 4$ (B) $y = 6$ (C) $y = 8$ (D) $y = 12$ (E) $y = \frac{48}{5}$

⑪ Which is equivalent to $(x - 5)(2x + 3)$?

(A) $2x^2 - 7x - 15$ (B) $2x^2 - 2x - 15$ (C) $x^2 - 2x - 15$

(D) $x^2 - 13x + 15$ (E) $2x^2 - 13x - 15$

⑫ Which is equivalent to $4x(2x^4 - 5)$?

(A) $8x^5 - 5x$ (B) $2x^4 - 20x$ (C) $8x^5 - 20$

(D) $8x^4 - 20x$ (E) $8x^5 - 20x$

⑬ Which is equivalent to $12x^5 - 16x^3$?

(A) $4x^2(3x^3 - 4)$ (B) $4x^2(3x^3 + 4)$ (C) $6x^2(2x^3 - 3x)$

(D) $4x^3(3x^2 - 4)$ (E) $4x^3(3x^2 + 4)$

⑭ What does $\sqrt{2x}$ times $\sqrt{8x}$ equal?

(A) $4\sqrt{x}$ (B) $2\sqrt{x}$ (C) $4x$ (D) $16x$ (E) $16x^2$

⑮ Solve for x in the equation $\frac{3}{x} = 9$.

(A) $x = \frac{1}{3}$ (B) $x = 6$ (C) $x = 3$ (D) $x = 12$ (E) $x = 27$

⑯ Solve for x in the equation $\frac{48}{x} = \frac{8}{3}$.

(A) $x = 6$ (B) $x = 16$ (C) $x = 18$ (D) $x = 128$ (E) $x = 144$

⑰ Which represents a length that is less than 5 meters?

(A) $L > -5$ (B) $L < 5$ (C) $L = 5$ (D) $L > 5$ (E) $5L > 1$

⑱ Isolate the variable in $6x - 9 < 9$.

(A) $x < -18$ (B) $x < -3$ (C) $x < 0$ (D) $x < 3$ (E) $x < 18$

⑲ Isolate the variable in $2 - 3x > 8$.

(A) $x < -2$ (B) $x < 2$ (C) $x < -\frac{10}{3}$ (D) $x > -2$ (E) $x > 2$

⑳ Which inequality is equivalent to $6 > -x$?

(A) $x < -6$ (B) $x < 6$ (C) $-6 > x$ (D) $x > 6$ (E) $x > -6$

-6-
RELATIONSHIPS

6.1 Additive relationships

6.2 Multiplicative relationships

6.3 Additive and multiplicative

6.4 Modeling simple relationships

6.5 Applying simple models

6.6 Linear relationships

6.7 Modeling linear relationships

6.8 Applying linear models

6.9 Direct and inverse proportions

6.10 Inverse relationships

6.11 Modeling inverse relationships

6.12 Applying inverse models

6.13 Properties of operators

6.14 Equivalent expressions

6.15 Modeling word problems

6.1 Additive relationships

In a simple additive relationship, two quantities differ by a fixed amount. For example, in the table below, Tom's age is always seven less than Sarah's age.

Tom	11	12	13	14	15
Sarah	18	19	20	21	22

Example. The table below shows gross and net golf scores. The top row shows gross scores. The bottom row subtracts 14 from the gross score: $86 - 14 = 72$, $90 - 14 = 76$, etc.

gross	86	90	94	98	102
net	72	76	80	84	88

Problem. ① Fill in the missing elements of the table.

Celsius	60	70			100
Kelvin		343	353	363	

Problem. ② Fill in the missing elements of the table.

price	4.2			5.1	
profit	1.8	2.1	2.4		3

Problem. ③ Fill in the missing elements of the table.

width	1/16	1/8	3/16	1/4	5/16
gap	5/16	3/8	7/16		

6.2 Multiplicative relationships

In a simple multiplicative relationship, multiply (or divide) one quantity by a fixed value to get the other quantity. For example, in the table below, multiply the number of yards by three to get the number of feet.

yards	2	3	4	5	6
feet	6	9	12	15	18

Example. The table below shows lengths in the top row. The bottom row multiplies the top row by four: $(12)(4) = 48$, etc.

length	12	15	18	21	24
area	48	60	72	84	96

Problem. ① Fill in the missing elements of the table.

distance	6	12	18	24	
time		144			360

Problem. ② Fill in the missing elements of the table.

Amps	3	4.5		7.5	9
Volts	7.2		14.4		

Problem. ③ Fill in the missing elements of the table.

force	1/72	1/36	1/24	1/18	5/72
work	1/18	1/9	1/6		

6.3 Additive and multiplicative

In this section, we will see whether you can tell if a given relationship is additive or multiplicative. If needed, review the tables from Sec.'s **6.1** and **6.2**.

Examples. The relationship below is additive because for each column, B is 8 more than A.

A	12	24	36	48	60
B	20	32	44	56	68

The relationship below is multiplicative because for each column, P is 5 times F.

F	12	24	36	48	60
P	60	120	180	240	300

Problem. ① Is the relationship additive or multiplicative?

f	1	2	3	4	5
E	2	4	6	8	10

Problem. ② Is the relationship additive or multiplicative?

w	4	8	12	16	20
z	8	12	16	20	24

Problem. ③ Is the relationship additive or multiplicative?

S	$8.25	$8.75	$9.25	$9.75	$10.25
E	$1.65	$1.75	$1.85	$1.95	$2.05

6.4 Modeling simple relationships

Multiplicative and additive relationships can be modeled using equations. For example, $P = 0.3V$ is a multiplicative relationship while $Q = D + 9$ is an additive relationship.

Examples. The relationship below can be modeled by $G = F + 16$ because each value of G is 16 more than F.

F	24	30	36	42	48
G	40	46	52	58	64

The relationship below can be modeled by $Q = 4T$ because each value of Q is 4 times T.

T	250	300	350	400	450
Q	1000	1200	1400	1600	1800

Problem. ① Write an equation to model the relationship.

p	45	65	85	105	125
L	180	260	340	420	500

Problem. ② Write an equation to model the relationship.

b	8	10	12	14	16
c	6	8	10	12	14

Problem. ③ Write an equation to model the relationship.

D	0.17	0.27	0.37	0.47	0.57
R	0.085	0.135	0.185	0.235	0.285

6.5 Applying simple models

The equation which models a relationship can be used to make predictions. For example, given that $P = 5T$, you can predict what P will be for a given value of T. For example, when $T = 16$, the model predicts that $P = 5(16) = 80$.

Examples. (A) Given $Z = A - 7$, predict Z when $A = 42$.

$$Z = A - 7 = 42 - 7 = 35$$

(B) Given $F = 3B$, predict F when $B = 15$.

$$F = 3B = 3(15) = 45$$

Problems. ① Given $U = 12C$, predict U when $C = 8$.

② Given $E = H + 1.2$, predict E when $H = 8.8$.

③ Given $T = 60F$, predict T when $F = 400$.

④ Given $j = k - \frac{1}{4}$, predict j when $k = \frac{7}{12}$.

⑤ Given $I = 0.15P$, predict I when $P = \$1200$.

6.6 Linear relationships

A linear relationship involves multiplication and addition. For example, in the table below, multiply the profit by 3 and add 2 to get the sales. For example, $6(3) + 2 = 18 + 2 = 20$.

profit	6	8	10	12	14
sales	20	26	32	38	44

Example. The table below shows pounds in the top row. The bottom row multiplies the top row by three and subtracts one: $3(3) - 1 = 8$, $6(3) - 1 = 17$, $9(3) - 1 = 26$, etc.

pounds	3	6	9	12	15
cost	8	17	26	35	44

Problem. ① Fill in the missing elements of the table.

red	8	12		20	24
green	34		66		98

Problem. ② Fill in the missing elements of the table.

students	150		200		
teachers	7	8	9	10	11

Problem. ③ Fill in the missing elements of the table.

weight	0.4	0.7			1.6
price	1.9		4.9	6.4	

6.7 Modeling linear relationships

Linear relationships have the form $y = mx + b$, where x and y are the variables and m and b are constants. For example, the table below can be modeled by $y = 4x + 2$. For example, $4(3) + 2 = 14$, $4(4) + 2 = 18$, and $4(5) + 2 = 22$.

x	3	4	5	6	7
y	14	18	22	26	30

You can find the coefficient using the formula $\frac{y_2 - y_1}{x_2 - x_1}$. For the table above, the coefficient is $\frac{y_2 - y_1}{x_2 - x_1} = \frac{18 - 14}{4 - 3} = \frac{4}{1} = 4$.

Example. Write an equation to model the relationship.

P	2	4	6	8	10
C	5	9	13	17	21

Subtract two consecutive C values and two consecutive P values, then divide. This forms the coefficient.

$$\frac{C_2 - C_1}{P_2 - P_1} = \frac{9 - 5}{4 - 2} = \frac{4}{2} = 2$$

If you multiply each value of P by this coefficient (2), you will still need to add 1 to make C. The relationship can thus be modeled by $\boxed{C = 2P + 1}$. Check: $2(2) + 1 = 5$, $2(4) + 1 = 9$, $2(6) + 1 = 13$, $2(8) + 1 = 17$, and $2(10) + 1 = 21$.

Problem. ① Write an equation to model the relationship.

t	0.6	0.9	1.2	1.5	1.8
v	1.5	2.1	2.7	3.3	3.9

Problem. ② Write an equation to model the relationship.

Q	11	15	19	23	27
R	75	103	131	159	187

Problem. ③ Write an equation to model the relationship.

T	40	80	120	160	200
U	215	455	695	935	1175

6.8 Applying linear models

A model for a relationship can be used to make predictions. If the variable you are predicting isn't already solved for, you may need to isolate the unknown (Sec.'s 5.8 to 5.9).

Examples. (A) Given $y = 9x - 3$, predict y when $x = 5$.

$$y = 9x - 3 = 9(5) - 3 = 45 - 3 = 42$$

(B) Given $H = 2G - 4$, predict G when $H = 6$.

First isolate G. Add 4 to get $H + 4 = 2G$, then divide by 2 to get $\frac{H+4}{2} = G$.

$$G = \frac{H + 4}{2} = \frac{6 + 4}{2} = \frac{10}{2} = 5$$

Problems. ① Given $P = 0.4S - 1.2$, predict P when $S = 7$.

② Given $u = 3t - 1$, predict t when $u = 14$.

③ Given $F = 1.8C + 32$, predict F when $C = 100$.

6.9 Direct and inverse proportions

In $g = 6f$, the variables f and g are directly proportional because an increase in f results in an increase in g. In $q = \frac{2}{p}$, the variables p and q are inversely proportional because an increase in p results in a smaller value of q. For example, $\frac{2}{9}$ is smaller than $\frac{2}{3}$ (since $\frac{2}{3} = \frac{2(3)}{3(3)} = \frac{6}{9} > \frac{2}{9}$).

Examples. (A) $d = 8t$ is a direct proportion: A larger value of t causes a larger value of d. For example, $8(9) > 8(2)$.

(B) $v = \frac{24}{t}$ is an inverse proportion: A larger value of t causes a smaller value of v. For example, $\frac{24}{8} < \frac{24}{4}$ since $3 < 6$.

Problems. Are you given a direct or inverse proportion?

① $Q = 9V$

② $P = \frac{120}{R}$

③ $M = \frac{30}{p}$

④ $v = \frac{f}{2}$

⑤ $W = 60F$

⑥ $\frac{1}{x} = \frac{1}{y}$

⑦ $nv = 3$

⑧ $\frac{V}{R} = 0.4$

6.10 Inverse relationships

In a simple inverse relationship, two quantities have a fixed product. For example, in the table below, a red value times a corresponding blue value is 240. Example: $2(120) = 240$.

red	2	3	4	5	6
blue	120	80	60	48	40

Example. In the table below, freq. times period equals one. For example, $2\left(\frac{1}{2}\right) = 1$. Put another way, period equals one divided by freq.: For example, when freq. $= 2$, period $= \frac{1}{2}$.

freq.	2	4	6	8	10
period	1/2	1/4	1/6	1/8	1/10

Problem. ① Fill in the missing elements of the table.

time	4	8	16	32	
rate	128	64			8

Problem. ② Fill in the missing elements of the table.

Amps			0.08	0.016	0.0032
Ohms	5	25	125		

Problem. ③ Fill in the missing elements of the table.

base	2			5	6
height		32	24		16

6.11 Modeling inverse relationships

Simple inverse relationships have the form $y = \frac{c}{x}$, where x and y are the variables and c is a constant. For example, the table below can be modeled by $y = \frac{120}{x}$ since $\frac{120}{2} = 60$, $\frac{120}{3} = 40$, $\frac{120}{4} = 30$, $\frac{120}{5} = 24$, and $\frac{120}{6} = 20$.

x	2	3	4	5	6
y	60	40	30	24	20

Example. The relationship below can be modeled by $a = \frac{320}{m}$ since $\frac{320}{4} = 80$, $\frac{320}{8} = 40$, $\frac{320}{16} = 20$, $\frac{320}{32} = 10$, and $\frac{320}{64} = 5$.

m	4	8	16	32	64
a	80	40	20	10	5

Problem. ① Write an equation to model the relationship.

d	5	10	20	40	80
V	800	400	200	100	50

Problem. ② Write an equation to model the relationship.

A	0.002	0.02	0.2	2	20
P	30,000	3000	300	30	3

Problem. ③ Write an equation to model the relationship.

L	560	480	400	320	240
W	1/7	1/6	1/5	1/4	1/3

6.12 Applying inverse models

If the variable you are predicting isn't already solved for, you may need to isolate the unknown (Sec.'s 5.8 to 5.9).

Examples. (A) Given $y = \frac{48}{x}$, predict y when $x = 6$.

$$y = \frac{48}{x} = \frac{48}{6} = 8$$

(B) Given $f = \frac{1}{T}$, predict T when $f = 25$.

First isolate T. Multiply by T to get $fT = 1$ then divide by f to get $T = \frac{1}{f}$.

$$T = \frac{1}{f} = \frac{1}{25} = \frac{1(4)}{25(4)} = \frac{4}{100} = 0.04$$

Problems. ① Given $I = \frac{2}{R}$, predict I when $R = 5$.

② Given $r = \frac{1200}{t}$, predict t when $r = 60$.

③ Given $bh = 3.6$, predict h when $b = 0.8$.

6.13 Properties of operators

Basic operator properties are summarized below.

Commutative Property	Associative Property
$x + y = y + x$ $xy = yx$	$(x + y) + z = x + (y + z)$ $(xy)z = x(yz)$
Distributive Property	**Identity Property**
$x(y + z) = xy + xz$ $x(y - z) = xy - xz$	$x + 0 = x$ $1x = x$
Inverse Property	
$x + (-x) = 0$	$x\left(\dfrac{1}{x}\right) = 1$

Examples. Which property is being applied?

(A) $\dfrac{3x}{3} = x$ uses the inverse property of multiplication.

(B) $3(2x - 8) = 6x - 24$ uses the distributive property.

Problems. Which properties are being applied?

① $-(1 - x) = -1 + x$ ② $-1 + x = x - 1$

③ $1 + x - 1 = x$ ④ $(x + 2)3 = 3x + 6$

⑤ $4(2 - x)2 = 8(2 - x)$ ⑥ $\dfrac{1}{2x}\dfrac{1}{3} = \dfrac{1}{6x}$

6.14 Equivalent expressions

The basic properties of operators can be applied to generate equivalent expressions. For example, when we combine like terms in $9x - 4x = 5x$, we use the distributive property in reverse (called factoring) to write $9x - 4x = x(9 - 4) = x(5)$ and then use the commutative property of multiplication to write $x(5) = (5)x = 5x$.

Examples. Are the given expressions equivalent?

(A) $3 + x - 3$ is equivalent to x since $3 - 3 = 0$.

(B) $2(x + 2)$ isn't equivalent to $2x + 1$ since $2(x + 2) = 2x + 4$.

Problems. Are the given expressions equivalent?

① $8x - 2 - 5x + 7$, $3x + 5$

② $\frac{1}{4}(4x - 1)$, $x - 1$

③ $2x - 3 + 4x - 5$, $2(3x - 4)$

④ $5x^2 - 4x + 3x^2$, $4x(2x - 1)$

6.15 Modeling word problems

One way to solve a word problem is to model the situation with an equation. For example, if Ed pays $60 for 4 chairs, by defining x to be the cost of each chair, we can model this situation as $4x = \$60$ to find that $x = \frac{\$60}{4} = \$15$ per chair.

Examples. Model the word problem and solve the equation.

(A) The total for Nichole's groceries comes to $11.73. Nichole pays with a $20 bill. How much money should she expect in return?

x = the amount of money Nichole should receive back

$$\$11.73 + x = \$20$$
$$x = \$20 - \$11.73$$
$$x = \$8.27$$

(B) Mrs. Higgins has 72 candies and 18 students. Mrs. Higgins wishes to distribute the candies evenly such that each student receives the same number of candies. How many candies should each student receive?

x = the number of candies each student should receive

$$18x = 72$$
$$x = \frac{72}{18} = 4$$

Problems. Model the word problem and solve the equation.

① After Sarah spent $14.28 on school supplies, she had $19.85 remaining. How much money did Sarah have to begin with?

② Jeff is 15 years old. Jeff is 5 times as old as Cindy. How old is Cindy?

③ Daniel wrote down two numbers on a piece of paper. One number is twice the other number. The sum of the numbers is 36. What are the numbers?

Multiple Choice Questions

① Which is next in the pattern $38, 45, 52, 59, 66$?

      (A) 69   (B) 71   (C) 72   (D) 73   (E) 74

② Which is next in the pattern $4, 8, 16, 32, 64$?

      (A) 80   (B) 96   (C) 128   (D) 256   (E) 512

③ What is the relationship between t and u below?

(A) additive   (B) multiplicative   (C) inverse   (D) other

t	4	6	8	10	12
u	12	18	24	30	36

④ What is the relationship between F and G below?

(A) additive   (B) multiplicative   (C) inverse   (D) other

F	3	9	15	21	27
G	6	12	18	24	30

⑤ Which models the relationship between p and q below?

  (A) $p = 2q$ (B) $q = 2p$ (C) $p = q + 12$ (D) $q = p + 12$

p	12	15	18	21	24
q	24	30	36	42	48

⑥ Which models the relationship between b and c below?

  (A) $b = \frac{4}{3}c$ (B) $c = \frac{4}{3}b$ (C) $b = c + 5$ (D) $c = b + 5$

b	15	30	45	60	75
c	20	35	50	65	80

⑦ Given $w = \frac{z}{2} - 3$, predict w when $z = 8$.

  (A) $w = 1$ (B) $w = 4$ (C) $w = 5$ (D) $w = 13$ (E) $w = 16$

⑧ Given $b = 3h + 1$, predict h when $b = 22$.

  (A) $h = 7$ (B) $h = \frac{22}{3}$ (C) $h = 21$ (D) $h = 66$ (E) $h = 67$

⑨ Which models the relationship between x and y below?

  (A) $y = x + 2$ (B) $y = x + 8$ (C) $y = x + 11$ (D) $y = 4x + 2$

x	3	5	7	9	11
y	14	22	30	38	46

⑩ In $xy = 5$, what is the relationship between x and y?

  (A) direct (B) inverse (C) associative (D) distributive

⑪ In $e = 3d$, what is the relationship between d and e?

  (A) direct (B) inverse (C) associative (D) distributive

⑫ Which property is applied in $x(x - 1) = x^2 - x$?

(A) associative (B) commutative (C) distributive (D) inverse

⑬ Which is equivalent to $9x + 1 - 7x - 7$?

(A) $4(x - 2)$ (B) $2(x + 3)$ (C) $2(x - 3)$ (D) $2(x + 4)$ (E) $2(x - 4)$

⑭ Sam spent $5 and has $9 left. Model this situation.

(A) $x - 5 = 9$ (B) $x + 5 = 9$ (C) $5x = 9$ (D) $9x = 5$ (E) $x + 9 = 5$

⑮ Pat paid $8 for 5 apples. Model this situation.

(A) $x - 5 = 8$ (B) $x + 5 = 8$ (C) $5x = 8$ (D) $8x = 5$ (E) $x + 8 = 5$

-7-

DATA ANALYSIS

7.1 Mean value

7.2 Median

7.3 Range

7.4 Standard deviation

7.5 Interquartile range

7.6 Box-and-whisker plots

7.7 Frequency

7.8 Dot plots

7.9 Histograms

7.10 Stem-and-leaf plots

7.11 Relative frequency and mode

7.12 Percent bar graphs

7.13 Pie charts

7.14 Data interpretation

7.1 Mean value

The mean value of a set of values equals the sum of the values divided by the number of values. The mean value is sometimes called the arithmetic mean or the average.

$$\text{mean} = \frac{V_1 + V_2 + V_3 + \cdots + V_N}{N}$$

The three dots (...) means "and so on," and N is the number of values. Check that your answer lies somewhere between the smallest and greatest values of the set.

Examples. Find the mean value.

(A) For 5 and 10, the mean is $\frac{5+10}{2} = \frac{15}{2} = \frac{15 \times 5}{2 \times 5} = \frac{75}{10} = 7.5.$

(B) For 2, 7, and 9, the mean is $\frac{2+7+9}{3} = \frac{18}{3} = 6.$

Problems. Find the mean value.

① 12, 13, 15, 16

② 4, 6, 7, 8, 9

7.2 Median

The median refers to the middle value. To find the median:

- If the set has an odd number of values, put the values in order. The median is the middle value.
- If the set has an even number of values, put the values in order. Find the mean of the two middle values.

Examples. Find the median.

(A) For 40, 29, and 38, the median is 38.

(B) For 74, 63, 89, and 70, the median is $\frac{70+74}{2} = \frac{144}{2} = 72$.

Problems. Find the median and the mean value.

① 16, 11, 9, 15, 14

② 8, 1, 4, 2, 9, 6

7.3 Range

The range equals the difference between the greatest value and the least value in a set of a data. The range provides a measure of the full spread of the data, whereas the median and mean provide measures of the center of the data.

Examples. Find the range.

(A) For 15, 12, 8, and 10, the range is $15 - 8 = 7$.

(B) For 34, 37, 33, 32, and 36, the range is $37 - 32 = 5$.

Problems. Find the range, median, and mean value.

① 13, 4, 9, 6, 8, 14

② 1.3, 0.5, 0.9, 1.1

7.4 Standard deviation

The standard deviation provides a useful measure of the spread of the data. The standard deviation is smaller than the range. The formula for the standard deviation is:

$$\text{std. dev.} = \sqrt{\frac{(V_1 - \text{mean})^2 + (V_2 - \text{mean})^2 + \cdots + (V_N - \text{mean})^2}{N-1}}$$

You need to find the mean before you can find the standard deviation. Note that N is the number of values in the set, and that the denominator $(N-1)$ is inside the square root.

Example. Find the standard deviation for 5, 7, and 9.

First find the mean value.

$$\text{mean} = \frac{5+7+9}{3} = \frac{21}{3} = 7$$

$$\text{std. dev.} = \sqrt{\frac{(5-7)^2 + (7-7)^2 + (9-7)^2}{3-1}}$$

$$\text{std. dev.} = \sqrt{\frac{(-2)^2 + (0)^2 + (2)^2}{2}} = \sqrt{\frac{4+0+4}{2}} = \sqrt{\frac{8}{2}} = \sqrt{4} = \boxed{2}$$

Note that $(-2)^2 = (-2) \times (-2) = 4$ (Sec. 1.3).

Problems. Find the mean, standard deviation, range, and median.

    ① 24, 27, 21

    ② 4, 10, 4, 5, 2

7.5 Interquartile range

The interquartile range (IQR) is an alternative measure of the spread of a data set. Find the IQR as follows:

- First divide the data into lower and upper halves.

- The lower quartile (LQ) is the median of the lower half.

- The upper quartile (UQ) is the median of the upper half.

- The interquartile range is: IQR = UQ − LQ.

Example. Find the IQR for 7, 3, 5, 9, 5, 4, 2, 8, 4, and 6.

The lower half includes 2, 3, 4, 4, and 5.

The upper half includes 5, 6, 7, 8, and 9.

The median of the lower half is LQ = 4.

The median of the upper half is UQ = 7.

The interquartile range is IQR = UQ − LQ = 7 − 4 = $\boxed{3}$.

Example. Find the IQR for 14, 10, 17, 9, 8, 16, 8, and 20.

The lower half includes 8, 8, 9, and 10.

The upper half includes 14, 16, 17, and 20.

The median of the lower half is LQ = $\frac{8+9}{2} = \frac{17}{2} = 8.5$.

The median of the upper half is UQ = $\frac{16+17}{2} = \frac{33}{2} = 16.5$.

The interquartile range is IQR = 16.5 − 8.5 = $\boxed{8}$.

Problems. Find the median, range, and interquartile range.

① $63, 47, 54, 48, 61, 51, 59, 50, 62, 54, 48, 59$

② $4.2, 3.6, 4.1, 3.8, 4, 4.1, 3.9, 4.2, 4.3, 3.9$

7.6 Box-and-whisker plots

A box-and-whisker plot shows the least value (LV), greatest value (GV), median (MD), lower quartile (LQ), upper quartile (UQ), and interquartile range (IQR). To make a box plot:

- Order the data from least to greatest.

- Find the LV, GV, MD, LQ, and UQ.

- Draw and label a number line that includes all of the data values. Look at the least and greatest values.

- Draw dots for the LV, GV, MD, LQ, and UQ.

- Draw a box with ends passing through the LQ and UQ.

- Add a vertical line through the MD.

- Draw whiskers. These are lines connecting the LV to the LQ and the UQ to the GV.

- The IQR extends from the LQ to the UQ: IQR = UQ − LQ.

Example. Make a box plot for 22, 25, 27, 29, 30, 32.

LV = 22, GV = 32, MD = $\frac{27+29}{2}$ = $\frac{56}{2}$ = 28, LQ = 25, UQ = 30

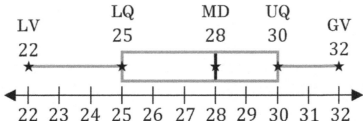

Problems. For each data set, make a box-and-whisker plot. Also determine the interquartile range.

① 16, 14, 17, 13, 15, 14, 16, 12

② 44, 42, 47, 44, 49, 42, 47, 50, 44, 44, 40, 49

7.7 Frequency

The frequency indicates how many times a particular data value appears in a set of data. For example, in 3, 3, 4, 4, 4, the frequency of 3's is 2 and the frequency of 4's is 3.

Example. Find the frequency of each data value below.

$$32, 34, 32, 31, 32, 33, 33, 31, 32, 30, 32, 32$$

Count each data value:

- The frequency of 30's is $\boxed{1}$ since 30 appears just once.
- The frequency of 31's is $\boxed{2}$ since 31 appears two times.
- The frequency of 32's is $\boxed{6}$ since 32 appears six times.
- The frequency of 33's is $\boxed{2}$ since 33 appears two times.
- The frequency of 34's is $\boxed{1}$ since 34 appears just once.

Tip: Add up the frequencies and check that it agrees with the total number of data values: $1 + 2 + 6 + 2 + 1 = 12$.

Problem. ① Find the frequency of each data value below.

$$7, 8, 7, 5, 7, 9, 6, 7, 8, 10, 7, 6, 8, 9$$

7.8 Dot plots

A dot plot organizes the data values by frequency. To make a dot plot:

- Draw and label a number line that includes all of the data values. Look at the least and greatest values.
- For each data value, place a dot (●) directly above its value on the number line.

Example. Make a dot plot for 3, 5, 3, 4, 3, 2, 4, 1, 2, and 3.

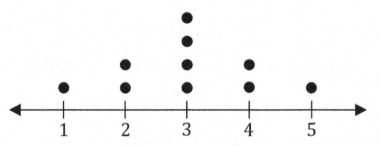

Problem. ① Make a dot plot for the data below.

5, 6, 4, 5, 7, 5, 4, 3, 5, 10, 6, 6, 8, 7

7.9 Histograms

A histogram divides the data into intervals and represents frequency with vertical bars. To make a histogram:

- Divide the data into intervals with equal size, such as 1-3, 4-6, 7-9, 10-12, and 13-15.

- Organize the data into a frequency table, showing the frequency for each interval.

- Draw a horizontal axis. Label the intervals here.

- Draw a vertical axis to represent frequency.

- Draw a vertical bar for each interval. The height of the bar corresponds to the frequency.

Example. Make a histogram for the data below.

5, 8, 4, 12, 5, 2, 6, 7, 4, 3, 15, 6, 10, 9

Int.	Freq.
1-3	2
4-6	6
7-9	3
10-12	2
13-15	1

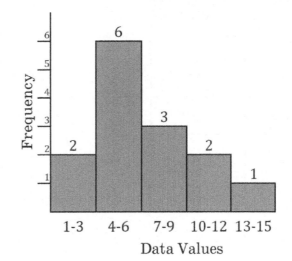

Problems. For each data set, complete the table and make a histogram.

① 8, 5, 7, 4, 8, 6, 9, 7, 10, 1, 8, 3, 6, 7, 5

Int.	Freq.
1-2	
3-4	
5-6	
7-8	
9-10	

② 11, 8, 14, 9, 3, 12, 20, 10, 15, 7, 9, 5, 16

Int.	Freq.
1-4	
5-8	
9-12	
13-16	
17-20	

7.10 Stem-and-leaf plots

Multi-digit numbers can be split into stems and leaves. For example, for the data set 37, 39, 42, 43, and 46, we can think of the tens digits as the stems and the units digits as the leaves. A stem-and-leaf plot organizes data by stems and leaves. To make a stem-and-leaf plot:

- Put the data into groups with the same stem (which can be the tens digit, for example).

- For each stem, write the data values in order by the leaves (which can be the units digits, for example).

Example. Make a stem-and-leaf plot for the data below.

41, 28, 47, 42, 42, 29, 40, 44, 24, 43

	Stems	Leaves
Put 24, 28, 29 here →	2	4 8 9
There aren't any 30's →	3	
Put 40, 41, 42, 42, 43, 44, 47 here →	4	0 1 2 2 3 4 7

Problem. ① Make a stem-and-leaf plot for the data below.

91, 83, 78, 95, 87, 95, 84, 89, 75, 97, 90, 77, 80

7.11 Relative frequency and mode

The relative frequency is the ratio of a particular frequency to the sum of the frequencies. Relative frequency may be expressed as a fraction or percent (review Sec.'s 3.7 to 3.9). The mode is the result that occurs most frequently. If two or more results tie for highest frequency, there can be more than one answer for the mode. (However, if *every* frequency is identical, then there isn't a mode.)

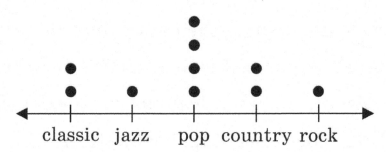

Example. Find the relative frequencies and the mode for the dot plot above.

There are $2 + 1 + 4 + 2 + 1 = 10$ data values.

Classic	Jazz	Pop	Country	Rock
2	1	4	2	1
$\frac{2}{10} = 20\%$	$\frac{1}{10} = 10\%$	$\frac{4}{10} = 40\%$	$\frac{2}{10} = 20\%$	$\frac{1}{10} = 10\%$

The mode is $\boxed{\text{pop}}$ since it has the highest frequency.

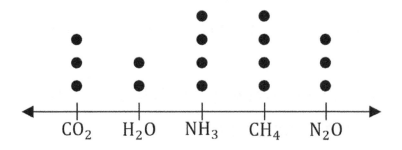

Problem. ① Find the relative frequencies for the dot plot above. Express them as percentages. Also find the mode(s).

7.12 Percent bar graphs

A percent bar graph helps to visualize relative frequency. To make a percent bar graph:

- First make a table of relative frequencies (Sec. 7.11). Tip: Check that the percent values add up to 100%.
- Draw a horizontal axis. Label the categories here.
- Draw a vertical axis to represent relative frequency.
- Draw a vertical bar for each interval. The height of the bar corresponds to the relative frequency. (Unlike a histogram, leave a gap between the bars.)

Red	Blue	Yellow	Green	Pink
15%	30%	25%	5%	25%

Example. Make a percent bar graph for the table above.

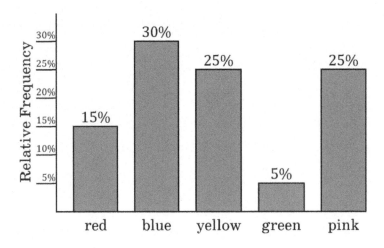

Pants	Shorts	Dress	Skirt	Other
40%	30%	15%	10%	5%

Problem. ① Make a percent bar graph for the table above. Also find the mode(s).

7.13 Pie charts

The relative frequency is indicated by the size of the pie slice.

Algebra	Geometry	Trig	Calculus
60%	20%	15%	5%

Example. Make a percent pie chart for the table above.

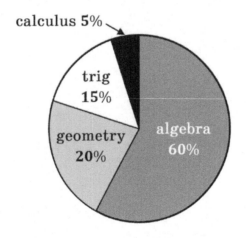

Problem. ① Make a percent pie chart for the table below.

Baseball	Soccer	Tennis	Golf
50%	25%	12.5%	12.5%

7.14 Data interpretation

Following are examples of quantitative measures that can be determined from a set of data:

- measures of the center: the mean value (or average) and the median
- measures of the spread: the range, the interquartile range, and the standard deviation
- measures of frequency: mode and relative frequency

Following are examples of qualitative characteristics that can be determined from a set of data:

- measures of position: peaks and clusters
- measure of shape: is it approximately symmetric?
- outliers: are any data points far from the others?

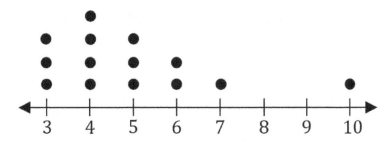

Example. The dot plot above is not symmetric, and the data points are clustered from 3 to 7 with a seeming outlier at 10. There is a peak at 4.

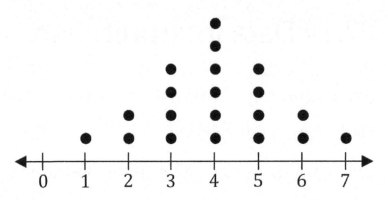

Problems. ① Determine the mean value, median, and range for the dot plot shown above.

② Describe the qualitative characteristics of the dot plot shown above, including peaks, clusters, shape, and outliers.

Multiple Choice Questions

Hours spent doing homework
2.1, 1.3, 0.8, 1.5, 2.3, 1.6

① What is the mean value for the table above?

    (A) 1.5   (B) 1.55   (C) 1.6   (D) 6   (E) 9.6

② What is the median for the table above?

    (A) 1.3   (B) 1.4   (C) 1.5   (D) 1.55   (E) 1.6

③ What is the range for the table above?

    (A) 0.4   (B) 0.8   (C) 1.3   (D) 1.5   (E) 6

④ What is the interquartile range for the table above?

    (A) 0.4   (B) 0.8   (C) 1   (D) 1.3   (E) 1.5

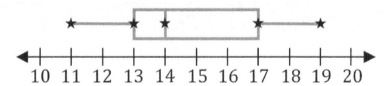

⑤ What is the median for the box plot above?

    (A) 13   (B) 14   (C) 14.8   (D) 15   (E) 17

⑥ What is the range for the box plot above?

    (A) 3   (B) 4   (C) 8   (D) 9   (E) 10

⑦ What is the interquartile range for the box plot above?

    (A) 1   (B) 2   (C) 3   (D) 4   (E) 8

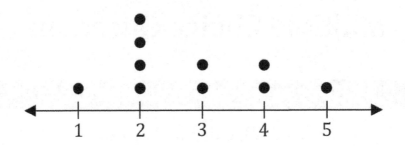

⑧ What is the mean value for the dot plot above?

(A) 2   (B) 2.5   (C) 2.8   (D) 3   (E) 5.6

⑨ What is the median for the dot plot above?

(A) 2   (B) 2.5   (C) 2.8   (D) 3   (E) 5.6

⑩ What is the range for the dot plot above?

(A) 1   (B) 2   (C) 3   (D) 4   (E) 5

⑪ What is the frequency of 2's for the dot plot above?

(A) 1   (B) 2   (C) 3   (D) 4   (E) 5

Pennies	Nickels	Dimes	Quarters
24	12	9	3

⑫ What is the mode for the table above?

(A) 3   (B) 24   (C) 50%   (D) pennies   (E) quarters

⑬ Find the relative frequency of nickels for the table above.

(A) 12%   (B) 25%   (C) 50%   (D) 200%   (E) 250%

⑭ What is the total amount of money for the table above?

(A) $0.48   (B) $2.49   (C) $2.64   (D) $4.29   (E) $7.68

Stems	Leaves
5	8
6	
7	2 5 5 8 9
8	0 3 4 4 7 9
9	1 1 6 8

⑮ Find the mean value for the stem-and-leaf plot above.

(A) 5   (B) 82.5   (C) 83   (D) 83.5   (E) 84

⑯ Find the median for the stem-and-leaf plot above.

(A) 5   (B) 82.5   (C) 83   (D) 83.5   (E) 84

⑰ Find the range for the steam-and-leaf plot above.

(A) 4   (B) 5   (C) 8   (D) 26   (E) 40

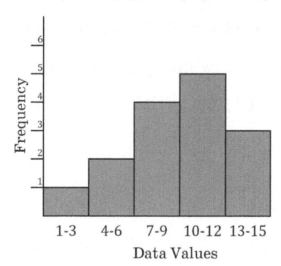

⑱ What is the frequency of data values in the range 7-9 for the histogram above?

(A) 1   (B) 2   (C) 3   (D) 4   (E) 5

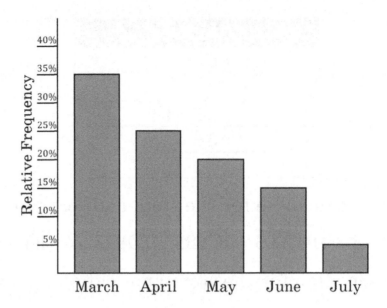

⑲ What is the mode for the percent bar graph above?

(A) 5%   (B) 35%   (C) 40%   (D) March   (E) July

⑳ The percent bar graph above shows data gathered from asking 80 people in which of these months they would prefer to take a vacation. How many people answered April?

(A) 16   (B) 20   (C) 24   (D) 25   (E) 28

-8-

COORDINATE GRAPHS

8.1 Ordered pairs

8.2 The four quadrants

8.3 Slope from coordinates

8.4 Slope from a graph

8.5 The y-intercept

8.6 Equation for a straight line

8.7 Equation from points

8.8 Equation from a table

8.9 Equation from a graph

8.10 Graphing an equation

8.1 Ordered pairs

An ordered pair has the form (x, y) and locates a point on the coordinate plane. The x-axis is horizontal with $+x$ to the right and the y-axis is vertical with $+y$ upward. The x- and y-axes intersect at the origin with coordinates $(0, 0)$. The first coordinate (x) indicates how far the point is to the right of the origin, while the second coordinate (y) indicates how far the point is above the origin.

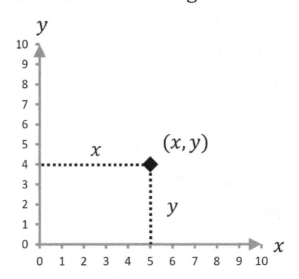

Example. Find the (x, y) coordinates of the point above.

The coordinates are $(5, 4)$ because the point ($\blacklozenge$) is $x = 5$ units to the right of the origin and $y = 4$ units above the origin.

Problems. Find the (x, y) coordinates of each point below.

① ②

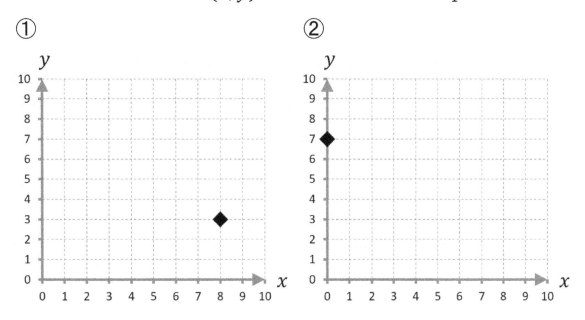

③ ④

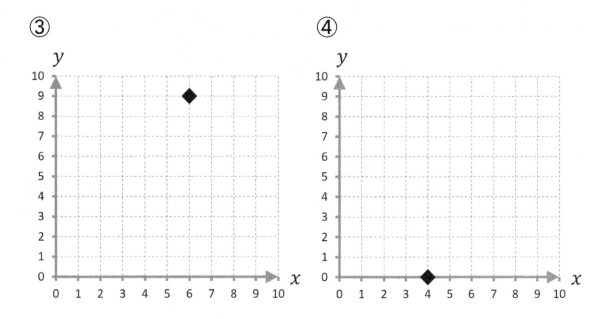

8.2 The four quadrants

The x- and y-axes divide the coordinate plane into four quadrants, labeled I thru IV as shown below. Note that left makes x negative while down makes y negative.

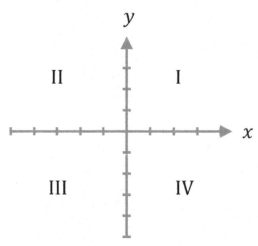

Examples. Find the (x, y) coordinates of each point below.

(A) y is down: $(2, -4)$ (B) x is left: $(-4, 2)$

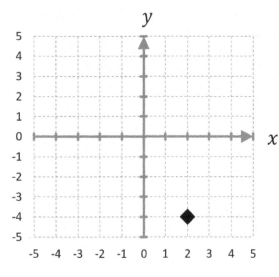

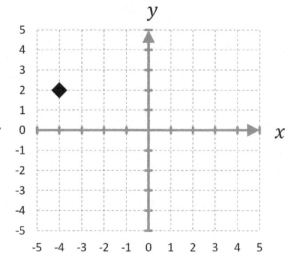

Problems. Find the (x, y) coordinates of each point below.

①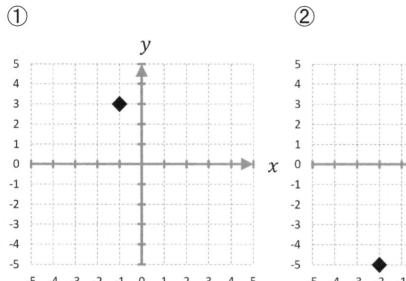

②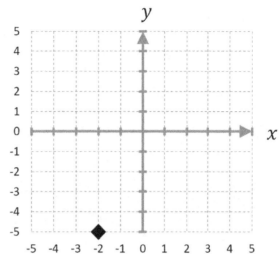

③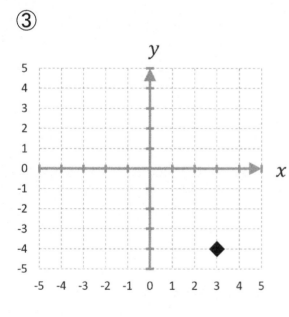

④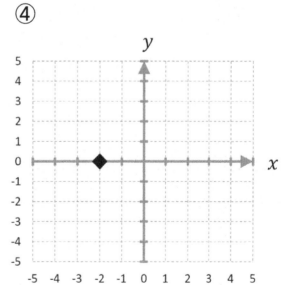

8.3 Slope from coordinates

The slope of a line indicates how steep it is. The greater the slope, the steeper the line. A horizontal line has zero slope. A positive slope angles up to the right, whereas a negative slope angles down to the right.

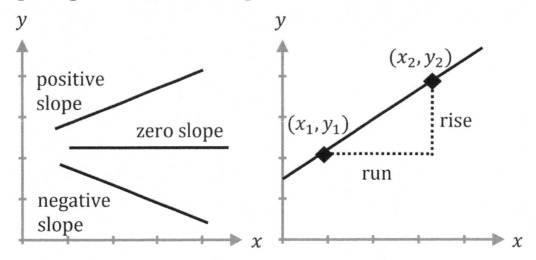

The slope of a line equals the rise over the run.

$$\text{slope} = \frac{\text{rise}}{\text{run}} = \frac{y_2 - y_1}{x_2 - x_1}$$

Examples. Find the slope of the line that passes through each pair of given points.

(A) $(2, 1)$ and $(4, 9)$: slope $= \frac{y_2 - y_1}{x_2 - x_1} = \frac{9-1}{4-2} = \frac{8}{2} = \boxed{4}$

Note that $x_1 = 2$, $y_1 = 1$, $x_2 = 4$, and $y_2 = 9$.

(B) $(-1, 6)$ and $(2, 0)$: slope $= \frac{y_2 - y_1}{x_2 - x_1} = \frac{0-6}{2-(-1)} = \frac{-6}{2+1} = \frac{-6}{3} = \boxed{-2}$

Problems. Find the slope of the line that passes through each pair of given points.

① $(4, 2)$ and $(6, 18)$

② $(-3, 4)$ and $(5, 8)$

③ $(-4, 8)$ and $(4, -8)$

④ $(-7, -5)$ and $(9, -5)$

8.4 Slope from a graph

To find the slope of a line, follow these steps:

- Mark two points that lie on the line. Choose points far apart to reduce interpolation error.
- Read off the (x, y) values of the two points.
- Use the slope formula (Sec. 8.3).

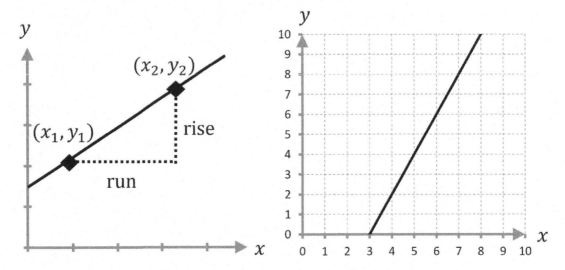

Example. Find the slope of the line above on the right.

The endpoints are $(3, 0)$ and $(8, 10)$.

Note that $x_1 = 3$, $y_1 = 0$, $x_2 = 8$, and $y_2 = 10$.

$$\text{slope} = \frac{y_2 - y_1}{x_2 - x_1} = \frac{10 - 0}{8 - 3} = \frac{10}{5} = \boxed{2}$$

Check: Every 1 unit horizontally the line rises 2 units vertically. The rise over the run is $\frac{2}{1} = 2$.

Problems. Find the slope of each line below.

①

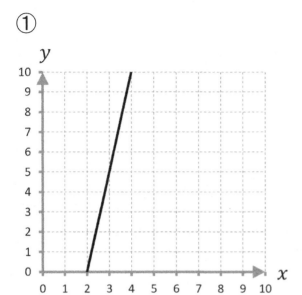

②

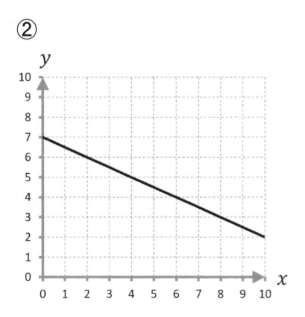

8.5 The *y*-intercept

The *y*-intercept is the value of *y* at the point where the line crosses the *y*-axis. **(Note that the *y*-axis is vertical.)**

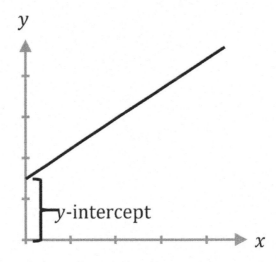

Examples. Find the *y*-intercept for each line below.

(A) The *y*-intercept is 3. **(B)** The *y*-intercept is −2.

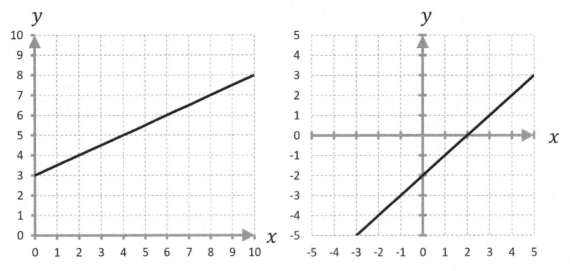

(Look where each line intersects the vertical axis.)

Problems. Find the *y*-intercept for each line below.

① ②

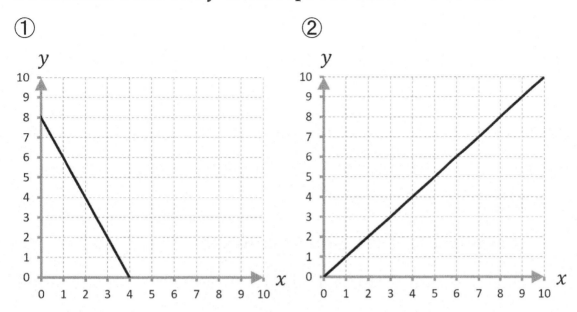

③ ④

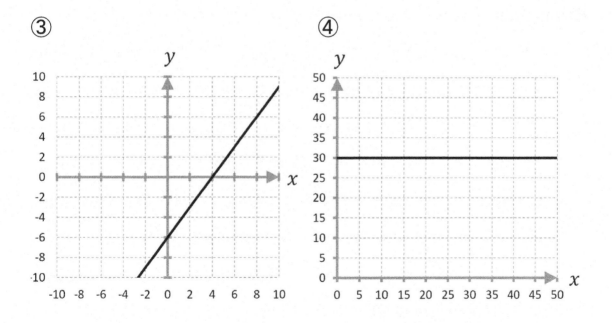

8.6 Equation for a straight line

The equation for a straight line can be put in the form:

$$y = mx + b$$

where m is the slope and b is the y-intercept. To determine the slope and y-intercept from the equation for a straight line, follow these steps:

- If applicable, first isolate y. Review Sec. 5.8.

- Once the equation is in the form $y = mx + b$, the slope is the coefficient of x and the y-intercept is added to the mx term. Note that m or b may be negative.

Examples. Find the slope and y-intercept of each line.

(A) $y = 5x - 7$: Compare $y = 5x - 7$ with $y = mx + b$ to see that the slope is $m = 5$ and the y-intercept is $b = -7$.

(B) $2y - 4x = 6$: First isolate y. Add $4x$ to both sides to get $2y = 4x + 6$. Divide both sides by 2 to get $y = 2x + 3$. Compare $y = 2x + 3$ with $y = mx + b$ to see that the slope is $m = 2$ and the y-intercept is $b = 3$.

(C) $y = -4$: Compare $y = -4$ with $y = mx + b$ to see that the slope is $m = 0$ and the y-intercept is $b = -4$.

Problems. Find the slope and y-intercept of each line.

① $y = x - 1$

② $\dfrac{y}{3} + 4x = 2$

③ $6 = 9x - 3y$

④ $x = -2y$

8.7 Equation from points

To determine the equation of a line that passes through two points, (x_1, y_1) and (x_2, y_2), follow these steps:

- First determine the slope (Sec. 8.3).
- Plug the x- and y-coordinates of either point into the equation $y = mx + b$ and isolate b (Sec. 5.8).
- Plug the values of m and b into $y = mx + b$.

Example. Determine the equation of the line that passes through $(1, 5)$ and $(3, 9)$.

First find the slope.

Note that $x_1 = 1$, $y_1 = 5$, $x_2 = 3$, and $y_2 = 9$.

$$m = \frac{y_2 - y_1}{x_2 - x_1} = \frac{9 - 5}{3 - 1} = \frac{4}{2} = 2$$

Plug $x = 1$, $y = 5$, and $m = 2$ into $y = mx + b$.

$$5 = 2(1) + b$$
$$5 = 2 + b$$
$$5 - 2 = b$$
$$3 = b$$

Plug $m = 2$ and $b = 3$ into $y = mx + b$. The equation for the line is $\boxed{y = 2x + 3}$.

Problems. Determine the equation of the line that passes through the given points.

① $(4, 3)$ and $(8, 5)$

② $(3, -4)$ and $(-2, 6)$

8.8 Equation from a table

Given a table of (x, y) coordinates that lie on one straight line, you just need one pair of (x, y) coordinates in order to determine the equation of the line (as outlined in Sec. 8.7). You can then use the equation for the line to predict other values of y for given values of x (or vice-versa).

x	2	4	6	
y	5	9		17

Example. For the table above, write an equation for the line and fill in the missing values of the table.

Two points are $(2, 5)$ and $(4, 9)$: First find the slope.

$$m = \frac{y_2 - y_1}{x_2 - x_1} = \frac{9 - 5}{4 - 2} = \frac{4}{2} = 2$$

Plug $x = 2$, $y = 5$, and $m = 2$ into $y = mx + b$.

$$5 = 2(2) + b \rightarrow 5 = 4 + b \rightarrow 5 - 4 = b \rightarrow 1 = b$$

The equation for the line is $\boxed{y = 2x + 1}$.

When $x = 6$, we get $y = 2(6) + 1 = 12 + 1 = 13$.

When $y = 17$, we get $17 = 2x + 1 \rightarrow 17 - 1 = 2x \rightarrow$

$16 = 2x \rightarrow \frac{16}{2} = 8 = x$.

x	2	4	6	8
y	5	9	13	17

Problems. For each table below, write an equation for the line and fill in the missing values of the table.

①

x	4	8	12	
y	1	3		7

②

x	−4	−2		2
y	−9	−3	3	

8.9 Equation from a graph

To determine the equation for a straight line from a graph, follow these steps:

- Find the slope according to Sec. 8.4.
- If possible, find the y-intercept according to Sec. 8.5. If not, plug (x, y) for a point into $y = mx + b$ and solve for b.
- Plug the values of m and b into $y = mx + b$.

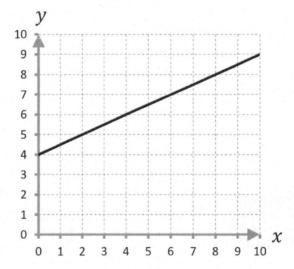

Example. Find the equation of the line above.

The endpoints are $(0, 4)$ and $(10, 9)$.

Note that $x_1 = 0$, $y_1 = 4$, $x_2 = 10$, and $y_2 = 9$.

$$m = \frac{y_2 - y_1}{x_2 - x_1} = \frac{9 - 4}{10 - 0} = \frac{5}{10} = \frac{1}{2} = 0.5$$

The y-intercept is 4. The equation is $\boxed{y = 0.5x + 4}$.

Problems. Find the equation of each line below.

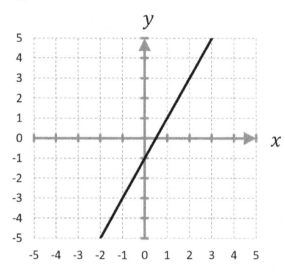

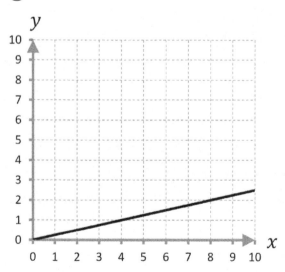

8.10 Graphing an equation

To draw a graph for the equation of a straight line, plug a variety of values of x into the equation, make a table of (x, y) values, plot the points, and draw a line through them.

Example. Graph the equation $y = 3x + 1$.

$$y = 3(0) + 1 = 0 + 1 = 1$$
$$y = 3(1) + 1 = 3 + 1 = 4$$
$$y = 3(2) + 1 = 6 + 1 = 7$$
$$y = 3(3) + 1 = 9 + 1 = 10$$

x	y
0	1
1	4
2	7
3	10

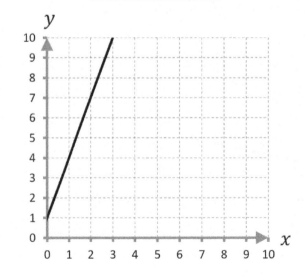

Problems. Graph each equation below.

① $y = -2x + 9$

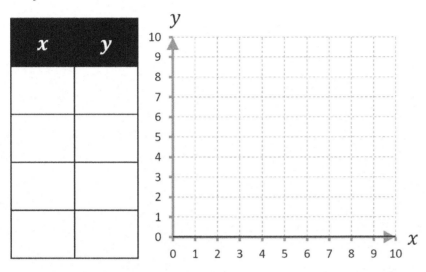

② $y = 4x - 8$

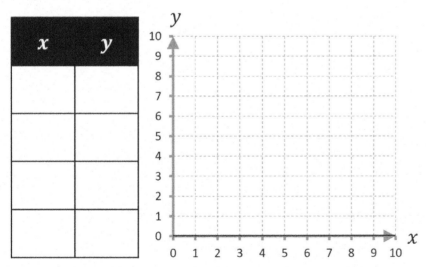

Multiple Choice Questions

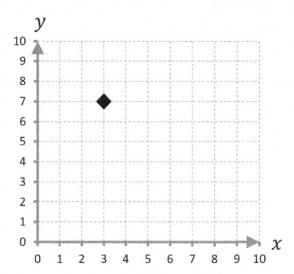

① What are the coordinates of the point shown above?

(A) $(0,3)$ (B) $(3,0)$ (C) $(3,7)$ (D) $(7,0)$ (E) $(7,3)$

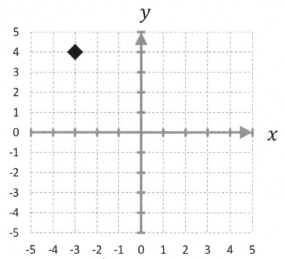

② In which quadrant is the point shown above?

(A) I   (B) II   (C) III   (D) IV

③ What are the coordinates of the point shown above?

(A) $(3,4)$ (B) $(-3,4)$ (C) $(3,-4)$ (D) $(-3,-4)$

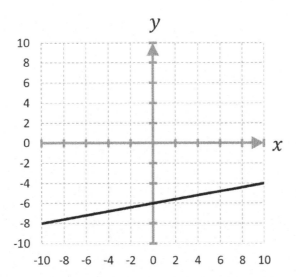

④ What is the slope of the line above?

(A) 0.2 (B) 0.4 (C) 2 (D) 2.5 (E) 5

⑤ What is the y-intercept of the line above?

(A) -8 (B) -6 (C) -4 (D) -3 (E) -2

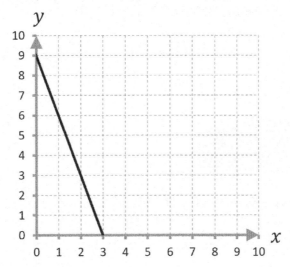

⑥ What is the equation for the line above?

(A) $y = -3x + 9$ (B) $y = 3x + 9$ (C) $y = -\frac{x}{3} + 9$

(D) $y = \frac{x}{3} + 9$ (E) $y = \frac{x}{9}$

⑦ What is the slope of the line connecting $(4, 2)$ and $(10, 20)$?

(A) $\frac{1}{6}$ (B) 0.5   (C) 1.8   (D) 2   (E) 3

⑧ What is the slope of $y = 0.2x - 8$?

(A) 0.2   (B) x (C) 5   (D) -8 (E) 8

⑨ What is the y-intercept of $y = 0.2x - 8$?

(A) 0.2   (B) x (C) 5   (D) -8 (E) 8

⑩ What is slope of $9x + 3y + 6 = 0$?

(A) -3 (B) -2 (C) 2   (D) 3   (E) 9

⑪ What is the y-intercept of $9x + 3y + 6 = 0$?

(A) -3 (B) -2 (C) 2   (D) 3   (E) 9

⑫ Which line passes through $(3, 8)$ and $(5, 14)$?

(A) $y = \frac{x}{3} - 7$ (B) $\frac{x}{3} + 7$ (C) $y = x + 5$

(D) $y = 3x - 1$ (E) $y = 3x + 1$

x	-2	1	4
y	5	-1	

⑬ Which line fits the data in the table above?

(A) $y = -2x + 1$ (B) $y = 0.5x + 1$ (C) $y = 0.5x - 1$

(D) $y = x + 7$ (E) $y = 2x + 1$

⑭ What is the missing value in the table above?

(A) -9 (B) -7 (C) 2   (D) 3   (E) 9

-9-

GEOMETRY

9.1 Unit conversions

9.2 Degrees and radians

9.3 Special angles

9.4 Triangles

9.5 Angle sum theorem

9.6 The triangle inequality

9.7 Pythagorean theorem

9.8 Corresponding parts

9.9 Quadrilaterals

9.10 Perimeter

9.11 Arc length

9.12 Area formulas

9.13 Volume formulas

9.14 Applying area and volume

9.1 Unit conversions

The idea behind unit conversions is to multiply by one in such a way as to cancel the old units and create new units. For example, since 1 yard = 3 feet, it follows that $\frac{1 \text{ yard}}{3 \text{ feet}} = 1$ and that $\frac{3 \text{ feet}}{1 \text{ yard}} = 1$ (since the numerator and denominator are equal). Multiplying by one doesn't change anything. The only issue is whether you want to multiply by $\frac{1 \text{ yard}}{3 \text{ feet}}$ or $\frac{3 \text{ feet}}{1 \text{ yard}}$. Choose the one that will cancel the old units:

$$12 \text{ feet} \times \frac{1 \text{ yard}}{3 \text{ feet}} = 4 \text{ yards}$$

$$4 \text{ yards} \times \frac{3 \text{ feet}}{1 \text{ yard}} = 12 \text{ feet}$$

Examples. (A) Convert 60 in. to feet given that 1 ft. = 12 in.

$$60 \text{ in.} \times \frac{1 \text{ ft.}}{12 \text{ in.}} = \frac{60}{12} = \boxed{5 \text{ ft.}}$$

(B) Convert 8 ft. to inches given that 1 ft. = 12 in.

$$8 \text{ ft.} \times \frac{12 \text{ in.}}{1 \text{ ft.}} = 8 \times 12 = \boxed{96 \text{ in.}}$$

(C) Convert 90 min. to hours given that 1 hr. = 60 min.

$$90 \text{ min.} \times \frac{1 \text{ hr.}}{60 \text{ min.}} = \frac{90}{60} = \frac{9}{6} = \frac{3}{2} = \boxed{1.5 \text{ hr.}}$$

Problems. ① Convert 7 cm to mm given that 1 cm = 10 mm.

② Convert 4 mm to cm given that 1 cm = 10 mm.

③ Convert 2.5 gal. to quarts given that 1 gal. = 4 qt.

④ Convert 375 g to kg given that 1 kg = 1000 g.

⑤ Convert 7.62 cm to inches given that 1 in. = 2.54 cm.

⑥ Convert 3 days to hours given that 1 day = 24 hr.

A rate like km/hr has a ratio of units. Write the units like fractions to help see the cancellation correctly.

Examples. (A) Convert 72 km/hr to m/s.

Note that 1 hr. = 60 min. = 3600 s.

$$72\,\frac{\cancel{km}}{\cancel{hr.}} \times \frac{1000\text{ m}}{1\,\cancel{km}} \times \frac{1\,\cancel{hr.}}{3600\text{ s}} = \frac{72{,}000}{3600} = \boxed{20\text{ m/s}}$$

(B) Convert 50 m/s to km/hr.

$$50\,\frac{\cancel{m}}{s} \times \frac{1\text{ km}}{1000\,\cancel{m}} \times \frac{3600\,\cancel{s}}{1\text{ hr.}} = \frac{180{,}000}{1000} = \boxed{180\text{ km/hr.}}$$

Note that $1\text{ yd.}^2 = 3\text{ ft.} \times 3\text{ ft.} = 9\text{ ft.}^2$ With units of area or volume, the conversion factor gets squared or cubed. In the diagram below, you can see visually that $1\text{ yd.}^2 = 9\text{ ft.}^2$

Examples. (C) Convert 4 yd.^2 to square feet.

$$4\text{ yd.}^2 \times \left(\frac{3\text{ ft.}}{1\text{ yd.}}\right)^2 = 4\,\cancel{\text{yd.}^2} \times \frac{9\text{ ft.}^2}{1\,\cancel{\text{yd.}^2}} = 4 \times 9 = \boxed{36\text{ ft.}^2}$$

(D) Convert 2 yd.^3 to cubic feet.

Note that $1\text{ yd.}^3 = 3\text{ ft.} \times 3\text{ ft.} \times 3\text{ ft.} = 27\text{ ft.}^3$

$$2\text{ yd.}^3 \times \left(\frac{3\text{ ft.}}{1\text{ yd.}}\right)^3 = 2\,\cancel{\text{yd.}^3} \times \frac{27\text{ ft.}^3}{1\,\cancel{\text{yd.}^3}} = 2 \times 27 = \boxed{54\text{ ft.}^3}$$

Problems. ① Convert 80 yd./min. to ft./s given that 1 yd. = 3 ft. and 1 min. = 60 s.

② Convert 36 in./s to ft./min. given that 1 ft. = 12 in. and 1 min. = 60 s.

③ Convert 0.24 cm² to mm² given that 1 cm = 10 mm.

④ Convert 3600 mm³ to cm³ given that 1 cm = 10 mm.

9.2 Degrees and radians

Angles can be measured in degrees (°) or radians (rad). In a full circle, there are 360° or 2π rad, where $\pi \approx 3.14159265$ is the Greek letter pi. The conversion factor is $180° = \pi$ rad.

Examples. (A) Convert 45° to radians.

$$45° \times \frac{\pi \text{ rad}}{180°} = \frac{45\pi}{180} = \boxed{\frac{\pi}{4} \text{ rad}}$$

(B) Convert $\frac{\pi}{3}$ rad to degrees.

$$\frac{\pi}{3} \text{rad} \times \frac{180°}{\pi \text{ rad}} = \frac{180\pi}{3\pi} = \frac{180}{3} = \boxed{60°}$$

Problems. ① Convert 120° to radians.

② Convert $\frac{\pi}{6}$ rad to degrees.

③ Convert $\frac{3\pi}{4}$ rad to degrees.

9.3 Special angles

Perpendicular lines form a 90° angle called a right angle. The symbol □ represents a right angle, and the symbol ⊥ means perpendicular.

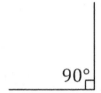

Complementary angles form a right angle; they add up to 90°. Supplementary angles form a straight line; they add up to 180°. The symbol ∠ indicates an angle.

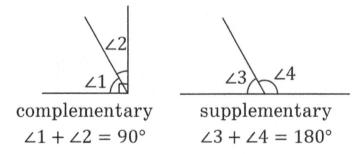

complementary
∠1 + ∠2 = 90°

supplementary
∠3 + ∠4 = 180°

Two lines intersect at a point called a vertex. Vertical angles form on opposite sides of a vertex. Vertical angles are equal.

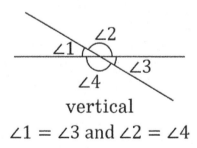

vertical
∠1 = ∠3 and ∠2 = ∠4

Example. Determine the unknown angle. (The diagram is not drawn to scale.)

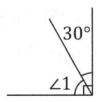

These angles are complementary; they add up to 90°.

$$\angle 1 + 30° = 90°$$

$$\angle 1 = 90° - 30° = \boxed{60°}$$

Problems. Determine each unknown angle. (The diagrams are not drawn to scale.)

①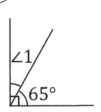

②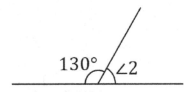

③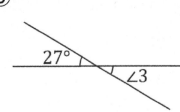

④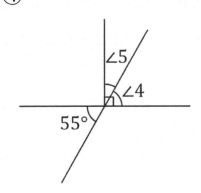

9.4 Triangles

An obtuse triangle has one angle greater than 90°, a right triangle has one 90° angle, and an acute triangle has three angles less than 90°.

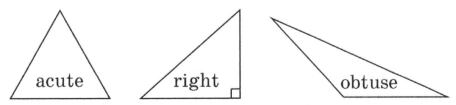

We use the word congruent to describe two sides that have the same length or two angles that have the same angular measure. An isosceles triangle has at least two congruent sides, an equilateral triangle has three congruent sides, and a scalene triangle doesn't have any congruent sides.

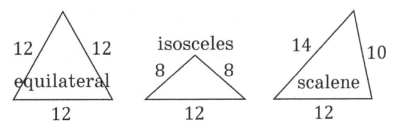

Problems. For each triangle, indicate if it is acute, right, obtuse, scalene, isosceles, and/or equilateral.

① ② ③

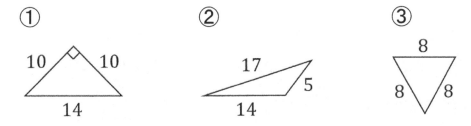

9.5 Angle sum theorem

The interior angles of any triangle add up to 180°.

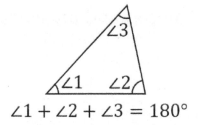

$$\angle 1 + \angle 2 + \angle 3 = 180°$$

Example. Determine the unknown angle.

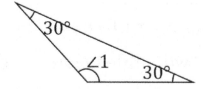

Apply the angle sum theorem.

$$\angle 1 + 30° + 30° = 180°$$

$$\angle 1 + 60° = 180°$$

$$\angle 1 = 180° - 60° = \boxed{120°}$$

Example. The three angles of an equilateral triangle are congruent. Show that they have an angular measure of 60°.

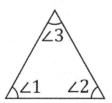

The angles are congruent: $\angle 1 = \angle 2 = \angle 3$. According to the angle sum theorem, $\angle 1 + \angle 1 + \angle 1 = 3(\angle 1) = 180°$, such that $\angle 1 = \frac{180°}{3} = \boxed{60°}$.

Problems. Determine each unknown angle. (The diagrams are not drawn to scale.)

①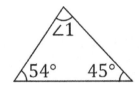

②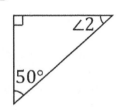

③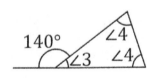

④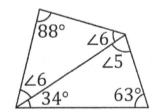

9.6 The triangle inequality

If you add up the lengths of any two sides of a triangle, that sum must be greater than the length of the remaining side. The diagram below shows that a triangle can't be formed when the triangle inequality isn't satisfied.

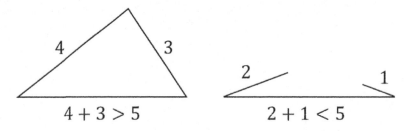

$$4 + 3 > 5 \qquad\qquad 2 + 1 < 5$$

Example. Is it possible to form a triangle with sides equal to 7 cm, 10 cm, and 2 cm?

Add up each pair of sides. It must be greater than the remaining side. Although $7 + 10 > 2$ and $10 + 2 > 7$, a triangle can't be formed because $7 + 2$ isn't greater than 10.

Example. Is it possible to form a triangle with sides equal to 5 in., 8 in., and 12 in.?

Add up each pair of sides. It must be greater than the remaining side. Since $5 + 8 > 12$, $5 + 12 > 8$, and $8 + 12 > 5$, a triangle can be formed.

Problems. For each set of lengths, determine whether or not it is possible to form a triangle.

① 7 m, 9 m, and 5 m

② 6 ft., 8 ft., and 1 ft.

③ 11 yd., 30 yd., and 18 yd.

④ 43 mm, 59 mm, and 17 mm

⑤ 2.3 mi., 3.5 mi., and 0.9 mi.

9.7 Pythagorean theorem

A right triangle has one 90° angle. The side opposite to the 90° angle is the longest side of the right triangle. It is called the hypotenuse. The shorter sides are called legs. According to the Pythagorean theorem, the sum of the squares of the legs of a right triangle equals the square of the hypotenuse.

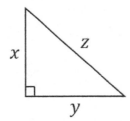 $x^2 + y^2 = z^2$

Examples. Find each unknown length.

(A)

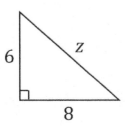

$6^2 + 8^2 = z^2$

$36 + 64 = z^2$

$100 = z^2$

$\sqrt{100} = \sqrt{z^2}$

$\boxed{10} = z$

(B)

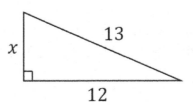

$x^2 + 12^2 = 13^2$

$x^2 = 169 - 144$

$x^2 = 25$

$\sqrt{x^2} = \sqrt{25}$

$x = \boxed{5}$

Problems. Determine each unknown length. (The diagrams are not drawn to scale.)

①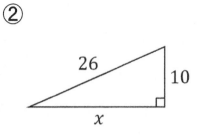

②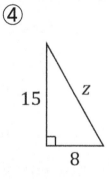

③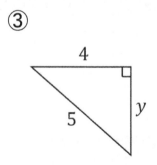

④

9.8 Corresponding parts

The shortest side of a triangle is opposite to the smallest angle, the middle side of a triangle is opposite to the middle angle, and the longest side of a triangle is opposite to the largest angle. For example, in the triangle below, $x < y < z$ and $\angle 1 < \angle 2 < \angle 3$.

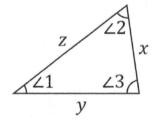

Example. The triangle below has sides with lengths of 10 cm, 17 cm, and 20 cm. Label these lengths on the triangle.

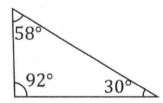

Label the shortest side (10 cm) opposite the smallest angle (30°) and the longest side (20 cm) opposite the largest angle (92°).

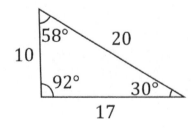

Problems. Label the given lengths on the appropriate sides of the triangle. (The diagrams are not drawn to scale.)

① 10 in., 11 in., and 18 in.   ② 5 m, 7 m, and 8 m

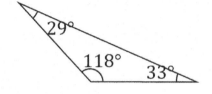   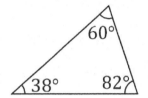

Problems. Label the given angles on the appropriate angles of the triangle. (The diagrams are not drawn to scale.)

③ 25°, 55°, and 100°   ④ 48°, 60°, and 72°

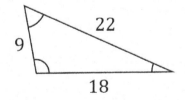   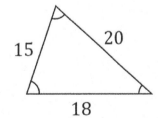

9.9 Quadrilaterals

A quadrilateral is a polygon with four sides. A trapezoid is a quadrilateral with one pair of parallel edges, while a parallelogram is a quadrilateral with two pairs of parallel edges. Special parallelograms include the rectangle (which has 90° angles), the rhombus (which has congruent edge lengths), and the square (90° angles and congruent edge lengths).

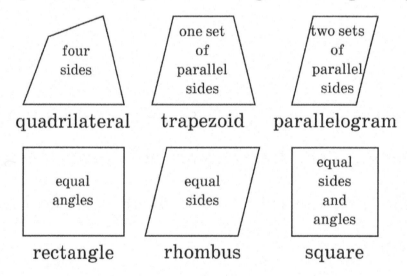

Problems. Give the most precise name of each shape below.

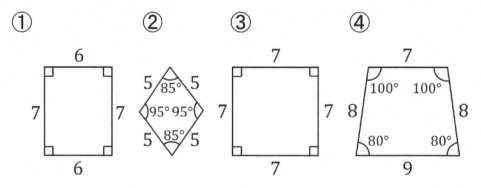

9.10 Perimeter

The perimeter of a polygon is the sum of its edge lengths.

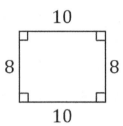

Example. Find the perimeter of the rectangle above.

$$P = 8 + 10 + 8 + 10 = 2(8) + 2(10) = 16 + 20 = 36$$

Problems. Find the perimeter of each polygon below.

① isosceles triangle

② trapezoid

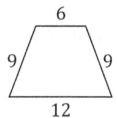

③ rectangle

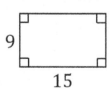

④ rhombus

9.11 Arc length

The length of a circular arc is $s = R\theta$, where s is called the arc length, R is the radius, and θ is the corresponding angle measured from the center of the circle. In the formula for arc length, θ must be expressed in radians (not degrees). To convert from degrees to radians, multiply by $\frac{\pi}{180}$ (Sec. 9.2). The arc length for a full circle is called circumference.

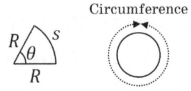

Example. Find the length of the arc below.

$$60°$$
$$6$$

First convert the angle to radians: $60° = 60° \times \frac{\pi \text{ rad}}{180°} = \frac{\pi}{3}$ rad.

The arc length is $s = R\theta = 6\left(\frac{\pi}{3}\right) = \boxed{2\pi} \approx 2(3.14) \approx \boxed{6.28}$.

Problems. Find the length of each arc below.

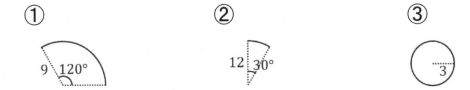

① 9 $120°$ ② 12 $30°$ ③ 3

9.12 Area formulas

The area of a rectangle equals its length times its width.

$$W \quad A = LW$$
$$L$$

Since a rectangle can be divided into two congruent right triangles, a right triangle has half the area of a rectangle.

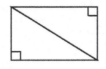

$$h \qquad A = \frac{1}{2}bh$$
$$b$$

As shown below, any triangle can be divided into two right triangles. Since $A_1 + A_2 = \frac{1}{2}b_1h + \frac{1}{2}b_2h = \frac{1}{2}(b_1 + b_2)h = \frac{1}{2}bh$, the area of any triangle is one-half times its base times its height. (Note that $b = b_1 + b_2$.)

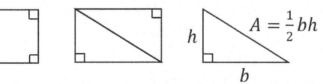

$$A = \frac{1}{2}bh \qquad h$$
$$b \qquad b_1 \quad b_2$$

The area of a trapezoid can be found by dividing the shape up into a rectangle and two triangles.

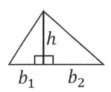

$$x + y + z = b_1$$

$$A = \frac{1}{2}xh + yh + \frac{1}{2}zh = \left(\frac{x}{2} + y + \frac{z}{2}\right)h$$

$$A = \frac{1}{2}(x + 2y + z)h = \frac{1}{2}(x + y + z + y)h$$

$$A = \frac{1}{2}(b_1 + b_2)h$$

The area of a parallelogram can be found by dividing it into two congruent triangles.

$$A = bh \qquad A = \tfrac{1}{2}bh + \tfrac{1}{2}bh = bh$$

The area of a rhombus can be found using the parallelogram formula or from half the product of its diagonals.

$$A = bh \qquad A = \tfrac{1}{2}d_1 d_2$$

Following are formulas for the area of a variety of common geometric shapes. Notes: d stands for "diagonal," and the diameter of a circle is twice its radius: $D = 2R$.

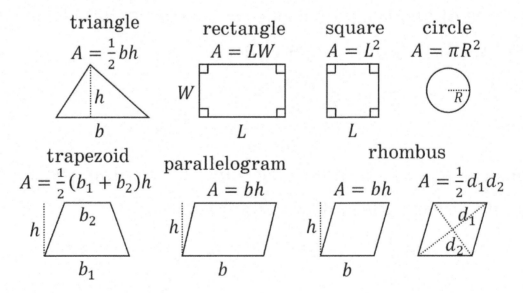

triangle
$$A = \tfrac{1}{2}bh$$

rectangle
$$A = LW$$

square
$$A = L^2$$

circle
$$A = \pi R^2$$

trapezoid
$$A = \tfrac{1}{2}(b_1 + b_2)h$$

parallelogram
$$A = bh$$

rhombus
$$A = bh \qquad A = \tfrac{1}{2}d_1 d_2$$

Example. Find the area of a parallelogram with a base of 3 in. and a height of 4 in.

$$A = bh = (3)(4) = 12 \text{ in.}^2$$

Problems. Find the area of each polygon below.

① triangle

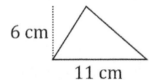

6 cm

11 cm

② a circle with a diameter of 24 m.

③ trapezoid

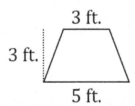

3 ft.

3 ft.

5 ft.

④ a rhombus with diagonals of 25 mm and 40 mm

9.13 Volume formulas

Following are formulas for the volume of a few common geometric shapes.

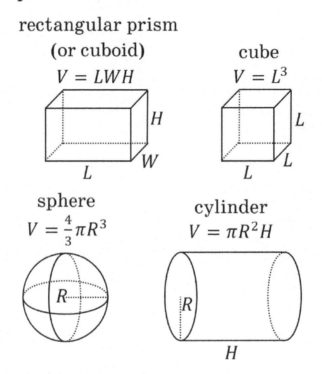

rectangular prism
(or cuboid)
$$V = LWH$$

cube
$$V = L^3$$

sphere
$$V = \frac{4}{3}\pi R^3$$

cylinder
$$V = \pi R^2 H$$

Example. What is the volume of a cube with 2-in. sides?

$$V = L^3 = 2^3 = 8 \text{ in.}^3$$

Problem. ① A rectangular prism has a length of 9 cm, a width of 5 cm, and a height of 4 cm. What is its volume?

9.14 Applying area and volume

Find area and volume formulas in Sec.'s 9.12 to 9.13.

Example. A parallelogram has a height of 2 m. What base does it need in order to have an area of 8 m^2?

Plug numbers into the relevant area equation.

$$A = bh$$
$$8 = b(2) = 2b$$

Isolate the unknown. Divide both sides by 2.

$$\frac{8}{2} = \boxed{4 \text{ m}} = b$$

Problem. ① A triangle has a base of 6 in. What height does it need in order to have an area of 24 in.^{2}?

Problem. ② What length does a square need in order to have an area of 9 ft.^{2}?

Problem. ③ What length does a cube need in order to have a volume of 64 cm^3?

Multiple Choice Questions

① Convert 78 in. to feet, given that 1 ft. = 12 in.

(A) 0.15 ft.   (B) 6 ft.   (C) 6.5 ft.   (D) 6.6 ft.   (E) 936 ft.

② Convert 18 yd.^{2} to ft.^{2}, given that 1 yd. = 3 ft.

(A) 2 ft.^{2} (B) 6 ft.^{2} (C) 54 ft.^{2} (D) 162 ft.^{2} (E) 324 ft.^{2}

③ Convert 1.5 m/s to km/hr., given that 1 km = 1000 m and 1 hr = 3600 s.

(A) 0.42 km/hr.   (B) 5.4 km/hr.   (C) 150 km/hr.

(D) 1500 km/hr.   (E) 8000 km/hr.

④ Convert 270° to radians.

(A) $\frac{\pi}{2}$ rad   (B) $\frac{\pi}{4}$ rad   (C) $\frac{2\pi}{3}$ rad   (D) $\frac{3\pi}{2}$ rad   (E) $\frac{3\pi}{4}$ rad

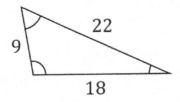

⑤ Which word best describes the triangle above?

(A) acute   (B) equilateral   (C) isosceles

(D) obtuse   (E) right

⑥ Which side could make a triangle with 5 and 8?

(A) 1   (B) 2   (C) 12   (D) 14   (E) 16

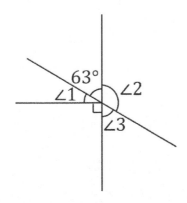

⑦ What is ∠1 in the diagram above?

(A) 27°   (B) 37°   (C) 63°   (D) 117°   (E) 153°

⑧ What is ∠2 in the diagram above?

(A) 27°   (B) 37°   (C) 63°   (D) 117°   (E) 153°

⑨ What is ∠3 in the diagram above?

(A) 27°   (B) 37°   (C) 63°   (D) 117°   (E) 153°

⑩ Which best describes a triangle with sides 6, 9, and x?

(A) $3 < x < 15$ (B) $3 < x < 9$ (C) $6 < x < 15$ (D) $6 < x$ (E) $x < 9$

⑪ A triangle has angles 42°, 71°, and x. What is x?

(A) 23°   (B) 29°   (C) 67°   (D) 113°   (E) 151°

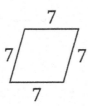

⑫ What is the most precise name of the shape above?

(A) parallelogram   (B) quadrilateral   (C) rectangle

(D) rhombus   (E) square

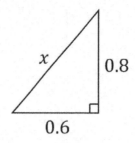

⑬ What is x in the triangle above?

(A) 0.53　(B) 0.7　(C) 1　(D) 1.4　(E) 3.16

⑭ A triangle has a base of 20 inches and a height of 25 inches. What is the area of the triangle?

(A) 25 in.^{2}　(B) 45 in.^{2}　(C) 90 in.^{2}　(D) 250 in.^{2}　(E) 500 in.^{2}

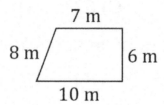

⑮ What is the perimeter of the shape shown above?

(A) 26 m　(B) 31 m　(C) 49 m　(D) 51 m　(E) 68 m

⑯ What is the area of the shape shown above?

(A) 26 m^2　(B) 31 m^2　(C) 49 m^2　(D) 51 m^2　(E) 68 m^2

⑰ A square has an area of 0.36 cm^2. What is the edge length of the square?

(A) 0.06 cm (B) 0.1296 cm (C) 0.6 cm (D) 1.296 cm (E) 6 cm

⑱ A rectangular prism has a volume of 60 ft.^{3} and a base with an area of 12 ft.^{2} What is its height?

(A) 0.2 ft. (B) 5 ft. (C) 6 ft. (D) 48 ft. (E) 720 ft.

-10-
FINANCE

10.1 Coins

10.2 Solving for money

10.3 Currency conversions

10.4 Sales tax

10.5 Simple interest

10.6 Bank accounts

10.7 Debits and credits

10.8 Balancing an account

10.9 Credit reports

10.1 Coins

A penny is worth 1 cent, a nickel is worth 5 cents, a dime is worth 10 cents, and a quarter is worth 25 cents. It takes 100 cents (¢) to make one dollar ($): 100¢ = $1.

Example. A boy has 7 quarters, 3 dimes, 6 nickels, and 14 pennies. What is the total value of this money in dollars?

$$7(25) + 3(10) + 6(5) + 14(1) = 175 + 30 + 30 + 14 = 249¢$$

$$\frac{249¢}{100} = \boxed{\$2.49}$$

Problem. ① A girl has 20 quarters, 8 dimes, 9 nickels, and 29 pennies. What is the total value of this money in dollars?

Problem. ② How can you make exactly $0.92 using as few coins as possible (without using half dollar coins)?

10.2 Solving for money

When working with money in decimal values, multiplying both sides of an equation by 100 removes the decimals. For example, $1.37x + 2.15 = 9$ becomes $137x + 215 = 900$.

Example. How many quarters make $2.75?

$$x = \text{ the number of quarters}$$
$$(\$0.25)x = \$2.75$$

Multiply both sides of the equation by 100.

$$25x = 275$$
$$x = \frac{275}{25} = \boxed{11}$$

Problem. ① How many nickels make $1.20?

Problem. ② A customer buys carrots and celery for $5.15. (There is no sales tax.) The carrots cost $2.85. How much did the celery cost?

10.3 Currency conversions

Currency conversions work just like unit conversions (Sec. 9.1). The difference is that the currency exchange rate can change. For example, one day a British pound (£) may be worth $1.25 and another day it might be worth $1.32.

Example. Convert £40 to dollars given that £1 = $1.25.

$$£40 \times \frac{\$1.25}{£1} = \boxed{\$50}$$

Problems. ① Convert $25 to pounds (£) given that £1 = $1.25.

② Convert 12€ (euro) to dollars given that 1€ = $1.1.

③ Convert ¥80 (yen) to dollars given that ¥1 = $0.009.

④ Convert $30 to rupees (Rs) given that Rs 1 = $0.015.

10.4 Sales tax

To figure the sales tax on a purchase, first convert the tax rate to a decimal (Sec. 3.7). Multiply the cost of the item(s) by the tax rate in decimal form (Sec. 3.6). To determine the total cost, add the value of the tax to the cost of the item(s).

Example. A girl buys a laptop for $400 where the sales tax is 8%. What is the total cost?

Convert the tax rate to a decimal: $8\% = \frac{8\%}{100\%} = 0.08$.

The tax is $400 \times 0.08 = \$32$.

The total cost is $400 + \$32 = \boxed{\$432}$.

Problems. ① A boy buys batteries for $11 where the sales tax is 9%. What is the total cost?

② A state charges 4% sales tax for groceries and 10% sales tax for everything else. A girl buys milk for $4 and a shirt for $12. What is the total cost?

10.5 Simple interest

The amount invested is called the principal (P). Convert the interest rate (r) to a decimal (Sec. 3.7), and multiply by the principal to determine the interest (I) earned: $I = Pr$.

Example. A woman invests \$600 in a savings account that earns interest at a rate of 3% per year. How much money will she have in her account after one year?

Convert the interest rate to a decimal: $3\% = \frac{3\%}{100\%} = 0.03$.

The interest earned is $I = Pr = \$600 \times 0.03 = \$18$.

The total amount is $\$600 + \$18 = \boxed{\$618}$.

Problems. ① A boy invests \$120 in a holiday fund that earns interest at a rate of 2%. How much will he earn in one year?

② A girl invests \$200 in an account that earns interest at a rate of 7% in one year. How much money will she have in her account after one year?

10.6 Bank accounts

Common types of bank accounts include:

- A checking account lets you make deposits, withdraw money, use a debit card, use an ATM machine, and pay by writing checks. Some banks charge more fees than other banks. One standard fee is to charge for using a debit card at another bank's ATM machine. A monthly fee for a low balance is sometimes waved when you set up direct deposit. You usually need to purchase checks.

- A savings account earns interest. There can be fees for a low balance or too many transactions. A CD usually offers a better interest rate than a savings account.

- Banks offer different types of loans, like a mortgage loan (for a home), a car loan, an unsecured loan, or a line of credit. Making a down payment can help reduce the interest paid and increase your approval chances.

- A credit card charges you interest (unless you pay the balance within the grace period). There are sometimes fees such as an annual fee or an over-the-limit fee.

Note the distinction between a debit card and credit card:

- A debit card works with a checking account. When you use the debit card, you are using money that you have in your checking account; you're not borrowing money from the bank. A few banks charge fees for using the debit card, most don't charge a fee for this, and a few banks actually pay you for each transaction. Using a debit card doesn't help you build a credit history.

- A credit card works like a loan. The bank is lending you money to make the purchase, and you must pay it back to the bank. If you don't pay the full amount within the grace period, you are charged interest. If you only make the minimum payments, it can take several years to finally get out of debt. Some credit cards give you rewards (like gift cards) for making purchases, but this is only a good value if you're not paying more money in interest and fees than you're receiving. By paying the balance in full each month within the grace period, you can avoid interest while also enjoying the rewards. Using a credit card responsibly helps to build a positive credit history.

Example. Bank A offers a credit card with a $50 annual fee and an interest rate of 15%, while Bank B offers a credit card with no annual fee and an interest rate of 20%. When would each credit card be the wiser choice?

If you plan to pay over time (instead of paying your balance in full each month) and have a balance of over $1000 on average (since $0.2 \times \$1000 - 0.15 \times \$1000 = \$200 - \$150 = \$50$), then Bank A's lower interest rate makes more sense. If you plan to pay your balance in full each month or have a balance of less than $1000 on average, then Bank B's zero annual fee is wiser.

Problem. ① Bank X charges a monthly fee of $10, but pays you $0.50 for every debit transaction you make, while Bank Y has no monthly fee, but doesn't pay for debit usage. When would each bank be the wiser choice?

10.7 Debits and credits

When keeping track of your bank account balance and when comparing with bank statements, a credit refers to money that goes into an account whereas a debit refers to money that comes out of an account.

Example. Is the action a debit or credit for your checking account?

(A) The bank charges you a monthly service fee.

This decreases your account balance. It is a debit.

(B) You deposit your weekly paycheck.

This increases your account balance. It is a credit.

Problems. Is the action a debit or credit for your checking account?

① You write a check to pay for groceries.

② You withdraw $20 from the ATM machine.

③ You deposit a roll of quarters from your piggy bank.

④ You pay for clothes using your debit card.

10.8 Balancing an account

The current balance is the amount of money currently in an account. The process of balancing an account helps you see what the current balance is. Add credits (like deposits) to the balance, but subtract debits (like withdrawals).

Example. The check register below shows a deposit in the credit column, and withdrawals and transfers in the debit column. Subtract each debit from the balance, and add each credit. Example: $1478.85 − $200.00 = $1278.85.

Check Number	Date	Transaction	Credit	Debit	Balance
	5/1	beginning balance			$1478.85
	5/1	ATM withdrawal		$200.00	$1278.85
	5/2	debit card for groceries		$52.41	$1226.44
	5/3	deposit from paycheck	$851.32		$2077.76
432	5/4	check electric bill		$145.24	$1932.52
	5/5	transfer to savings		$350.00	$1582.52
433	5/6	check car insurance		$215.48	$1367.04

Problem. ① Balance the check register below by filling in the rightmost column (called Balance).

Check Number	Date	Transaction	Credit	Debit	Balance
	7/1	beginning balance			$1832.27
517	7/1	check rent		$550.00	
	7/3	deposit from paycheck	$743.85		
	7/4	ATM withdrawal		$300.00	
	7/7	deposit birthday cash	$150.00		
518	7/9	check car payment		$362.75	
	7/11	debit card car repairs		$261.38	

10.9 Credit reports

Once you begin to pay bills and borrow money, you begin to establish a credit history. The information in your credit history appears in your credit report. Landlords and banks use credit reports to help determine whether or not to rent an apartment or approve a loan application. A credit score based on your credit report aids with this decision. A score of 700 indicates good credit. Paying your bills on time and keeping your credit card and loan balances low help to earn a better credit score. Negative information can remain on your credit report for seven years.

Problem. ① Would lenders prefer for you to have a high or low debt-to-income ratio? Explain.

Multiple Choice Questions

① What is the total value of 18 quarters, 7 dimes, 8 nickels, and 17 pennies?

(A) $0.50   (B) $4.77   (C) $4.82   (D) $5.77   (E) $5.82

② How many quarters make $6.25?

(A) 21   (B) 25   (C) 75   (D) 125   (E) 15,625

③ A bag of apples costs $3.29. A customer buys a bag of apples and a bag of oranges for $6.18. If there isn't any sales tax, how much does a bag of oranges cost?

(A) $2.89   (B) $2.91   (C) $3.89   (D) $3.91   (E) $9.47

④ Convert $30 to pounds (£) given that £1 = $1.25.

(A) £24   (B) £25   (C) £32.50   (D) £36   (E) £37.50

⑤ Convert £45 to dollars given that £1 = $1.25.

(A) $33.75   (B) $36   (C) $40   (D) $54   (E) $56.25

⑥ What is the tax on a $42 purchase with an 8% sales tax?

(A) $1.60   (B) $3.20   (C) $3.36   (D) $5.25   (E) $336

⑦ What does a $275 t.v. cost with 10% sales tax?

(A) $27.50   (B) $247.50   (C) $250   (D) $300   (E) $302.50

⑧ If you invest $400 in a savings account that earns 2.5% interest per year, how much do you earn in one year?

(A) $2.50   (B) $5   (C) $10   (D) $25   (E) $100

⑨ If you invest $2400 and earn 5% interest per year, what will the balance be after one year?

(A) $2280   (B) $2350   (C) $2405   (D) $2450   (E) $2520

⑩ A bank charges a $6 monthly service fee, but pays you $0.40 for each debit transaction that you make. How many debit transactions do you need to make in one month in order to offset the cost of the monthly service fee?

(A) 3   (B) 7   (C) 12   (D) 13   (E) 15

⑪ Which of the following would be a credit when balancing your checking account?

(A) payroll deposit   (B) ATM withdrawal

(C) debit card purchase   (D) writing a check

⑫ Bob's checking account balance is currently $632.85. Bob writes a check for $47.93 to pay for car repairs. What is his new checking account balance?

(A) $474.92 (B) $484.92 (C) $584.92 (D) $594.92 (E) $680.78

⑬ Taylor earns $360 per week at her job. She puts 25% of her earnings in a savings account. How much does she save each week?

(A) $72   (B) $90   (C) $180   (D) $450   (E) $480

⑭ Which of the following has a positive impact on your credit score?

(A) high credit card balance   (B) losing your job

(C) late payments   (D) low debt-to-income ratio

ANSWER KEY

Chapter 1: Arithmetic Operations

Page 6

① 13   ② 15   ③ 17
④ 16   ⑤ 16   ⑥ 14
⑦ 14   ⑧ 15   ⑨ 13
⑩ 14   ⑪ 12   ⑫ 18
⑬ 9   ⑭ 5   ⑮ 8
⑯ 7   ⑰ 9   ⑱ 7
⑲ 7   ⑳ 8   ㉑ 8
㉒ 8   ㉓ 7   ㉔ 9

Page 7

① 25   ② 48   ③ 45
④ 35   ⑤ 63   ⑥ 49
⑦ 36   ⑧ 40   ⑨ 54
⑩ 56   ⑪ 81   ⑫ 30
⑬ 72   ⑭ 42   ⑮ 64
⑯ 9   ⑰ 8   ⑱ 9
⑲ 6   ⑳ 6   ㉑ 6
㉒ 7   ㉓ 6   ㉔ 7
㉕ 7   ㉖ 9   ㉗ 9
㉘ 9   ㉙ 7   ㉚ 8

Page 8

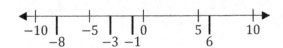

Page 9

    ① 7   ② 11

    ③ −9   ④ −1

    ⑤ −5   ⑥ −11

Page 10

    ① −15   ② −3

    ③ 16   ④ 6

    ⑤ −28   ⑥ −6

    ⑦ 3   ⑧ 2

    ⑨ −6   ⑩ −8

    ⑪ 1   ⑫ −1

Page 11

    ① 8   ② 9   ③ 1

    ④ 12   ⑤ 216   ⑥ 1

    ⑦ −64   ⑧ 1,000,000   ⑨ 81

    Note: $(-4)^3 = (-4) \times (-4) \times (-4) = 16 \times (-4) = -64$.

    Note: $(-3)^4 = (-3) \times (-3) \times (-3) \times (-3) = 9 \times 9 = 81$.

Page 12

    ① 20   ② 20   ③ 70

    ④ 100   ⑤ 10   ⑥ 0

    Note: 96 falls between 90 and 100. Since 6 is 5 or more, 96 rounds up to 100.

    Note: 4 and 5 fall between 0 and 10. Since 4 is less than 5, 4 rounds down to 0. Since 5 is 5 or more, 5 rounds up to 10.

    ⑦ 200   ⑧ 300   ⑨ 1000

    Note: 972 falls between 900 and 1000. Since 7 is 5 or more, 972 rounds up to 1000.

Page 13

    ① 5   ② 24   ③ 72

④ $-6 > -8$ ⑤ $-9 < 2$ ⑥ $-8 > -9$

⑦ $-3, -1, 2$ ⑧ $-31, -29, -20$

Page 14

① 21 ② 16 ③ 8

④ -16 ⑤ 4 ⑥ 25

⑦ 9 ⑧ 56 ⑨ 24

Page 15

① the commutative property of multiplication

② the distributive property

③ the identity property of multiplication

④ the commutative property of addition

Note that $5 - 3 = -3 + 5$ can be written as $5 + (-3) = (-3) + 5$, which is "addition" with the numbers 5 and negative 3. (Adding a negative number equates to subtracting a positive number.)

Page 16

There aren't any problems on page 16.

Page 17

① $(6 + 8 \div 2) \times 4 = (6 + 4) \times 4 = 10 \times 4 = 40$

Note: Parentheses first. Inside the parentheses, divide first then add.

② $30 - (9 - 6)^3 \div 3 = 30 - 3^3 \div 3 = 30 - 27 \div 3 = 30 - 9 = 21$

Note: Parentheses, exponent, divide, subtract.

③ $11 - 3 \times 2 + 8 \div 2^2 = 11 - 3 \times 2 + 8 \div 4 = 11 - 6 + 2 = 5 + 2 = 7$

Note: Exponent first. Multiply/divide left to right. Add/subtract left to right.

④ $(12 - 7)^2 - 9 \times 2 = 5^2 - 9 \times 2 = 25 - 18 = 7$

Note: Parentheses, exponent, multiply, subtract.

⑤ $15 - \frac{(9+7)}{(3-1)} = 15 - \frac{16}{2} = 15 - 8 = 7$

Note: Parentheses first. Then divide, then subtract.

⑥ $2 + (3 - 2^3) = 2 + (3 - 8) = 2 + (-5) = -3$

Note: Parentheses first. Inside the parentheses, exponent first then subtract.

⑦ $(9 - 4 \times 3)^3 \div 3 = (9 - 12)^3 \div 3 = (-3)^3 \div 3 = -27 \div 3 = -9$

Note: Parentheses first. Inside the parentheses, multiply first then subtract.

Note: $(-3)^3 = (-3) \times (-3) \times (-3) = 9 \times (-3) = -27$

⑧ $(4 - 3^2) - 15 \div 5 = (4 - 9) - 15 \div 5 = -5 - 15 \div 5 = -5 - 3 = -8$

Note: Parentheses first. Inside the parentheses, exponent first.

Note: It may help to review Sec. 1.3.

⑨ $\frac{(6+6)}{(1+2)} - \frac{(15+9)}{(5-3)} = \frac{12}{3} - \frac{24}{2} = 4 - 12 = -8$

Note: Parentheses, divide, subtract.

Page 18

① $4 \times 13 = 4 \times (10 + 3) = 4 \times 10 + 4 \times 3 = 40 + 12 = 52$

② $9 \times 72 = 9 \times (70 + 2) = 9 \times 70 + 9 \times 2 = 630 + 18 = 648$

③ $3 \times 99 = 3 \times (100 - 1) = 3 \times 100 - 3 \times 1 = 300 - 3 = 297$

Note that you could alternatively write $99 = 90 + 9$.

④ $45 \times 8 = (40 + 5) \times 8 = 40 \times 8 + 5 \times 8 = 320 + 40 = 360$

Page 19

There aren't any problems on page 19.

Page 20

① 2, 3, 6, 9   ② 3, 5

③ 2, 4   ④ 2, 3, 4, 6

⑤ 3   ⑥ 2, 5

⑦ 2, 3, 6   ⑧ 2, 3, 6, 9

⑨ 3, 5, 9   ⑩ 2, 3, 4, 6

⑪ 2, 3, 4, 5, 6, 9   ⑫ 2, 3, 4, 6, 9

Page 21

There aren't any problems on page 21.

Page 22

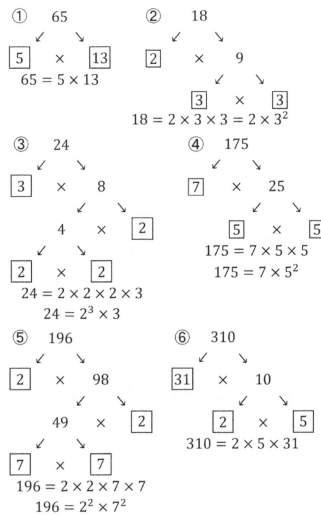

① 65

⑤ × ⑬
$65 = 5 \times 13$

② 18

② × 9

③ × ③
$18 = 2 \times 3 \times 3 = 2 \times 3^2$

③ 24

③ × 8

4 × ②

② × ②
$24 = 2 \times 2 \times 2 \times 3$
$24 = 2^3 \times 3$

④ 175

⑦ × 25

⑤ × ⑤
$175 = 7 \times 5 \times 5$
$175 = 7 \times 5^2$

⑤ 196

② × 98

49 × ②

⑦ × ⑦
$196 = 2 \times 2 \times 7 \times 7$
$196 = 2^2 \times 7^2$

⑥ 310

㉛ × 10

② × ⑤
$310 = 2 \times 5 \times 31$

Page 23

①

7	63	$63 \div 7 = 9$
3	9	$9 \div 3 = 3$
	3	3 is prime

$63 = 3 \times 3 \times 7$
$63 = 3^2 \times 7$

②

2	100	$100 \div 2 = 50$
2	50	$50 \div 2 = 25$
5	25	$25 \div 5 = 5$
	5	5 is prime

$60 = 2 \times 2 \times 5 \times 5$
$60 = 2^2 \times 5^2$

③

2	68	$68 \div 2 = 34$
2	34	$34 \div 2 = 17$
	17	17 is prime

$68 = 2 \times 2 \times 17$
$68 = 2^2 \times 17$

Page 24

    ① 6 $(2 \times 6 = 12, 3 \times 6 = 18)$ ② 9 $(3 \times 9 = 27, 4 \times 9 = 36)$

    ③ 22 $(1 \times 22 = 22, 3 \times 22 = 66)$ ④ 15 $(3 \times 15 = 45, 5 \times 15 = 75)$

    ⑤ 8 $(6 \times 8 = 48, 7 \times 8 = 56)$ ⑥ 6 $(4 \times 6 = 24, 6 \times 6 = 36, 7 \times 6 = 42)$

Page 25

    ① $6 + 15 = 3 \times 2 + 3 \times 5 = 3 \times (2 + 5)$

    ② $56 + 70 = 14 \times 4 + 14 \times 5 = 14 \times (4 + 5)$

    ③ $12 + 72 = 12 \times 1 + 12 \times 6 = 12 \times (1 + 6)$

    ④ $36 + 60 + 72 = 12 \times 3 + 12 \times 5 + 12 \times 6 = 12 \times (3 + 5 + 6)$

Page 26

    ① 5, 10, 15, 20, 25, $\boxed{30}$, 35 ... and 6, 12, 18, 24, $\boxed{30}$, 36 ... The LCM is 30.

    ② 8, 16, $\boxed{24}$, 32 ... and 12, $\boxed{24}$, 36 ... The LCM is 24.

    ③ 15, 30, 45, $\boxed{60}$, 75 ... and 20, 40, $\boxed{60}$, 80 ... The LCM is 60.

    ④ 21, 42, $\boxed{63}$, 84 ... and $\boxed{63}$, 126 ... The LCM is 63 (since 63 is a multiple of 21).

    ⑤ 16, 32, 48, 64, 80, 96, 112, 128, $\boxed{144}$ ... and 18, 36, 54, 72, 90, 108, 126, $\boxed{144}$ ...
The LCM is 144.

    ⑥ 12, 24, 36, 48, 60, 72, 84, $\boxed{96}$ ... , 16, 32, 48, 64, 80, $\boxed{96}$ ... , and 32, 64, $\boxed{96}$ ...
The LCM is 96.

Page 27

    ① 5   ② 8   ③ 7

    ④ 6   ⑤ 12   ⑥ 1

    ⑦ 0   ⑧ 10   ⑨ 16

Page 28

    ① $\sqrt{18} = \sqrt{9 \times 2} = \sqrt{9}\sqrt{2} = 3\sqrt{2}$

    ② $\sqrt{48} = \sqrt{16 \times 3} = \sqrt{16}\sqrt{3} = 4\sqrt{3}$

    ③ $\sqrt{28} = \sqrt{4 \times 7} = \sqrt{4}\sqrt{7} = 2\sqrt{7}$

    ④ $\sqrt{108} = \sqrt{36 \times 3} = \sqrt{36}\sqrt{3} = 6\sqrt{3}$

Page 29

    ① (C) $-16°C$

    ② (B) -8

    ③ (D) $8 - (-6)$ since $8 - (-6) = 8 + 6 = 14$

    ④ (A) -25 since it finished 25 below where it started

    ⑤ (B) -8 since $-12 - (-4) = -12 + 4 = -8$

    ⑥ (A) $(-8)^3$

    ⑦ (E) 625 since $(-5)^4 = (-5) \times (-5) \times (-5) \times (-5) = 25 \times 25$

    ⑧ (B) $(-4)^3$ since $(-4)^3 = (-4) \times (-4) \times (-4) = -64$ (sign matters)

    ⑨ (A) $-1, 1, |-2|, 3$

Note that $|-2| = 2$.

Page 30

    ⑩ (C) $6 \div 2$

Note: Parentheses first. Inside the parentheses, divide first.

    ⑪ (B) -19 since $\frac{24-6}{6-3} - (8 - 6 \div 2)^2 = \frac{18}{3} - (8 - 3)^2 = 6 - 5^2 = 6 - 25$

    ⑫ (C) $9 - (4 + 5)$ since $9 - (4 + 5) = 9 - 4 - 5 = 0$

    ⑬ (D) $5 \times 9 + 5 \times 2$ since $5 \times (9 - 2) = 5 \times 9 - 5 \times 2 = 45 - 10 = 35$

    ⑭ (A) 1980

    ⑮ (C) $2^3 \times 3^2$ since $72 = 8 \times 9 = 2^3 \times 3^2$

    ⑯ (E) 143 since $143 = 13 \times 11$

    ⑰ (C) 16 since $48 = 16 \times 3$ and $80 = 16 \times 5$

    ⑱ (B) 175 since $25 \times 7 = 175$ and $35 \times 5 = 175$

    ⑲ (A) $\sqrt{49}, 4^2, 5^2, 2^5$ since $\sqrt{49} = 7$, $4^2 = 16$, $5^2 = 25$, and $2^5 = 32$

Chapter 2: Fractions

Page 31

There aren't any problems on page 31.

Page 32

① $\frac{16}{20} = \frac{16 \div 4}{20 \div 4} = \frac{4}{5}$

② $\frac{13}{26} = \frac{13 \div 13}{13 \div 13} = \frac{1}{2}$

③ $\frac{15}{10} = \frac{15 \div 5}{10 \div 5} = \frac{3}{2}$

④ $\frac{18}{54} = \frac{18 \div 18}{54 \div 18} = \frac{1}{3}$

⑤ $\frac{96}{32} = \frac{96 \div 32}{32 \div 32} = \frac{3}{1} = 3$

⑥ $\frac{49}{28} = \frac{49 \div 7}{28 \div 7} = \frac{7}{4}$

Note: $\frac{3}{1}$ is the same as $3 \div 1$ (review Sec. 1.7).

Page 33

① $5\frac{2}{3} = \frac{5 \times 3 + 2}{3} = \frac{17}{3}$

② $6\frac{2}{5} = \frac{6 \times 5 + 2}{5} = \frac{32}{5}$

③ $1\frac{1}{2} = \frac{1 \times 2 + 1}{2} = \frac{3}{2}$

④ $10\frac{3}{4} = \frac{10 \times 4 + 3}{4} = \frac{43}{4}$

Page 34

① $\frac{27}{5} = 5\frac{2}{5}$ since $5 \times 5 = 25$ and $27 - 25 = 2$

② $\frac{10}{3} = 3\frac{1}{3}$ since $3 \times 3 = 9$ and $10 - 9 = 1$

③ $\frac{15}{2} = 7\frac{1}{2}$ since $7 \times 2 = 14$ and $15 - 14 = 1$

④ $\frac{29}{8} = 3\frac{5}{8}$ since $3 \times 8 = 24$ and $29 - 24 = 5$

Page 35

There aren't any problems on page 35.

Page 36

① $\frac{3}{8} = \frac{3 \times 3}{8 \times 3} = \boxed{\frac{9}{24}}$ and $\frac{7}{12} = \frac{7 \times 2}{12 \times 2} = \boxed{\frac{14}{24}}$

② $\frac{1}{15} = \frac{1 \times 5}{15 \times 5} = \boxed{\frac{5}{75}}$ and $\frac{2}{25} = \frac{2 \times 3}{25 \times 3} = \boxed{\frac{6}{75}}$

③ $\frac{5}{2} = \frac{5 \times 3}{2 \times 3} = \boxed{\frac{15}{6}}$ and $\frac{11}{6} = \frac{11 \times 1}{6 \times 1} = \boxed{\frac{11}{6}}$

④ $\frac{5}{24} = \frac{5 \times 4}{24 \times 4} = \boxed{\frac{20}{96}}$ and $\frac{7}{32} = \frac{7 \times 3}{32 \times 3} = \boxed{\frac{21}{96}}$

⑤ $\frac{1}{21} = \frac{1 \times 5}{21 \times 5} = \boxed{\frac{5}{105}}$ and $\frac{2}{35} = \frac{2 \times 3}{35 \times 3} = \boxed{\frac{6}{105}}$

⑥ $\frac{4}{3} = \frac{4 \times 5}{3 \times 5} = \boxed{\frac{20}{15}}$ and $\frac{8}{5} = \frac{8 \times 3}{5 \times 3} = \boxed{\frac{24}{15}}$

Page 37

There aren't any problems on page 37.

Page 38

① $\frac{5}{6} > \frac{7}{9}$ because $\frac{5}{6} = \frac{5 \times 3}{6 \times 3} = \frac{15}{18}, \frac{7}{9} = \frac{7 \times 2}{9 \times 2} = \frac{14}{18}$, and $\frac{15}{18} > \frac{14}{18}$

② $\frac{7}{12} < \frac{13}{20}$ because $\frac{7}{12} = \frac{7 \times 5}{12 \times 5} = \frac{35}{60}, \frac{13}{20} = \frac{13 \times 3}{20 \times 3} = \frac{39}{60}$, and $\frac{35}{60} < \frac{39}{60}$

③ $6 = \frac{72}{12}$ because $6 = \frac{6}{1} = \frac{6 \times 12}{1 \times 12} = \frac{72}{12}$ or because $\frac{72}{12} = 72 \div 12 = 6$

④ $\frac{11}{36} < \frac{17}{48}$ because $\frac{11}{36} = \frac{11 \times 4}{36 \times 4} = \frac{44}{144}, \frac{17}{48} = \frac{17 \times 3}{48 \times 3} = \frac{51}{144}$, and $\frac{44}{144} < \frac{51}{144}$

⑤ $\frac{2}{11} > \frac{5}{33}$ because $\frac{2}{11} = \frac{2 \times 3}{11 \times 3} = \frac{6}{33}, \frac{5}{33} = \frac{5 \times 1}{33 \times 1} = \frac{5}{33}$, and $\frac{6}{33} > \frac{5}{33}$

⑥ $4\frac{1}{3} < 5\frac{8}{9}$ because the mixed number $4\frac{1}{3}$ is less than 5, whereas the mixed number $5\frac{8}{9}$ is greater than 5. It may help to review Sec. 2.2.

Page 39

There aren't any problems on page 39.

Page 40

① $\frac{3}{4} + \frac{5}{6} = \frac{3 \times 3}{4 \times 3} + \frac{5 \times 2}{6 \times 2} = \frac{9}{12} + \frac{10}{12} = \frac{9 + 10}{12} = \frac{19}{12}$ alternate answer: $1\frac{7}{12}$

② $\frac{7}{4} - \frac{1}{12} = \frac{7 \times 3}{4 \times 3} - \frac{1 \times 1}{12 \times 1} = \frac{21}{12} - \frac{1}{12} = \frac{21 - 1}{12} = \frac{20}{12} = \frac{20 \div 4}{12 \div 4} = \frac{5}{3}$ alternate answer: $1\frac{2}{3}$

③ $\frac{27}{5} - 3 = \frac{27}{5} - \frac{3}{1} = \frac{27}{5} - \frac{3 \times 5}{1 \times 5} = \frac{27}{5} - \frac{15}{5} = \frac{27 - 15}{5} = \frac{12}{5}$ alternate answer: $2\frac{2}{5}$

④ $\frac{11}{30} + \left(-\frac{13}{45}\right) = \frac{11 \times 3}{30 \times 3} + \frac{-13 \times 2}{45 \times 2} = \frac{33}{90} + \frac{-26}{90} = \frac{33 + (-26)}{90} = \frac{7}{90}$

⑤ $-\frac{3}{14} - \frac{2}{35} = -\frac{3 \times 5}{14 \times 5} - \frac{2 \times 2}{35 \times 2} = -\frac{15}{70} - \frac{4}{70} = \frac{-15 - 4}{70} = \frac{-19}{70}$ alternate answer $-\frac{19}{70}$

⑥ $6\frac{1}{3} - 3\frac{4}{9} = \frac{19}{3} - \frac{31}{9} = \frac{19 \times 3}{3 \times 3} - \frac{31 \times 1}{9 \times 1} = \frac{57}{9} - \frac{31}{9} = \frac{26}{9}$ alternate answer $2\frac{8}{9}$

Page 41

① The reciprocal of $\frac{3}{4}$ equals $\frac{4}{3}$. Alternate answer: $1\frac{1}{3}$.

② The reciprocal of $\frac{10}{3}$ equals $\frac{3}{10}$.

③ The reciprocal of 4 equals $\frac{1}{4}$ (because $4 = 4 \div 1 = \frac{4}{1}$).

④ The reciprocal of $2\frac{2}{3}$ equals $\frac{3}{8}$ (because $2\frac{2}{3} = \frac{2\times3+2}{3} = \frac{8}{3}$).

⑤ The reciprocal of $\frac{1}{5}$ equals 5 (because $\frac{5}{1} = 5$).

⑥ The reciprocal of $7\frac{4}{5}$ equals $\frac{5}{39}$ (because $7\frac{4}{5} = \frac{7\times5+4}{5} = \frac{39}{5}$).

⑦ The reciprocal of 1 equals 1 (because $1 = 1 \div 1 = \frac{1}{1}$).

⑧ The reciprocal of $\frac{77}{8}$ equals $\frac{8}{77}$.

⑨ The reciprocal of $3\frac{1}{3}$ equals $\frac{3}{10}$ (because $3\frac{1}{3} = \frac{3\times3+1}{3} = \frac{10}{3}$).

Page 42

There aren't any problems on page 42.

Page 43

① $\frac{7}{2} \times \frac{5}{6} = \frac{7\times5}{2\times6} = \frac{35}{12}$ (alternate answer: $2\frac{11}{12}$)

② $\frac{3}{8} \times \frac{4}{9} = \frac{3\times4}{8\times9} = \frac{12}{72} = \frac{12\div12}{72\div12} = \frac{1}{6}$ (alternative: $\frac{3\times4}{8\times9} = \frac{3}{9} \times \frac{4}{8} = \frac{1}{3} \times \frac{1}{2} = \frac{1}{6}$)

③ $\frac{5}{12} \times 9 = \frac{5}{12} \times \frac{9}{1} = \frac{5\times9}{12\times1} = \frac{45}{12} = \frac{45\div3}{12\div3} = \frac{15}{4}$ (alternate answer: $3\frac{3}{4}$)

④ $1\frac{3}{4} \times \left(-3\frac{2}{3}\right) = \frac{7}{4} \times \left(-\frac{11}{3}\right) = \frac{7\times(-11)}{4\times3} = -\frac{77}{12}$ (alternate answer: $-6\frac{5}{12}$)

⑤ $\frac{1}{3} \div \frac{2}{7} = \frac{1}{3} \times \frac{7}{2} = \frac{1\times7}{3\times2} = \frac{7}{6}$ (alternate answer: $1\frac{1}{6}$)

⑥ $\frac{3}{4} \div \frac{9}{2} = \frac{3}{4} \times \frac{2}{9} = \frac{3\times2}{4\times9} = \frac{6}{36} = \frac{6\div6}{36\div6} = \frac{1}{6}$ (alternative: $\frac{3\times2}{4\times9} = \frac{3}{9} \times \frac{2}{4} = \frac{1}{3} \times \frac{1}{2} = \frac{1}{6}$)

⑦ $4 \div \frac{2}{3} = \frac{4}{1} \times \frac{3}{2} = \frac{4\times3}{1\times2} = \frac{12}{2} = 12 \div 2 = 6$

⑧ $\left(-4\frac{2}{9}\right) \div 2\frac{2}{3} = \left(-\frac{38}{9}\right) \div \frac{8}{3} = \left(-\frac{38}{9}\right) \times \frac{3}{8} = \frac{-38\times3}{9\times8} = -\frac{114}{72} = -\frac{114\div6}{72\div6} = -\frac{19}{12}$

(alternate answer: $-1\frac{7}{12}$)

⑨ $\left(-6\frac{1}{4}\right) \div \left(-\frac{3}{2}\right) = \left(-\frac{25}{4}\right) \times \left(-\frac{2}{3}\right) = \frac{-25\times(-2)}{4\times3} = \frac{50}{12} = \frac{50\div2}{12\div2} = \frac{25}{6}$ or $4\frac{1}{6}$

Note that $-25 \times (-2) = 50$. The two minus signs make a plus sign (Sec. 1.3).

Page 44

① $\left(\frac{8}{7}\right)^2 = \frac{8\times 8}{7\times 7} = \frac{64}{49}$ ② $\left(\frac{3}{4}\right)^5 = \frac{3\times3\times3\times3\times3}{4\times4\times4\times4\times4} = \frac{243}{1024}$

③ $\left(-2\frac{1}{2}\right)^3 = \left(-\frac{5}{2}\right)^3 = \frac{(-5)\times(-5)\times(-5)}{2\times2\times2} = -\frac{125}{8}$ (alternate answer: $-15\frac{5}{8}$)

④ $\left(-\frac{1}{2}\right)^8 = \frac{(-1)\times(-1)\times(-1)\times(-1)\times(-1)\times(-1)\times(-1)\times(-1)}{2\times2\times2\times2\times2\times2\times2\times2} = \frac{1}{256}$

Page 45

① $\left(\frac{5}{7}\right)^{-1} = \frac{7}{5}$ or $1\frac{2}{5}$ ② $\left(4\frac{1}{3}\right)^{-1} = \left(\frac{13}{3}\right)^{-1} = \frac{3}{13}$

③ $5^{-3} = \frac{1}{5^3} = \frac{1}{5\times5\times5} = \frac{1}{125}$ ④ $\left(\frac{4}{3}\right)^{-3} = \left(\frac{3}{4}\right)^3 = \frac{3\times3\times3}{4\times4\times4} = \frac{27}{64}$

⑤ $\left(3\frac{1}{3}\right)^{-5} = \left(\frac{10}{3}\right)^{-5} = \left(\frac{3}{10}\right)^5 = \frac{3\times3\times3\times3\times3}{10\times10\times10\times10\times10} = \frac{243}{100,000}$

Page 46

① $\frac{1}{\sqrt{2}} = \frac{1\times\sqrt{2}}{\sqrt{2}\times\sqrt{2}} = \frac{\sqrt{2}}{2}$ ② $\frac{7}{\sqrt{7}} = \frac{7\times\sqrt{7}}{\sqrt{7}\times\sqrt{7}} = \frac{7\sqrt{7}}{7} = \sqrt{7}$

③ $\frac{3}{\sqrt{11}} = \frac{3\times\sqrt{11}}{\sqrt{11}\times\sqrt{11}} = \frac{3\sqrt{11}}{11}$ ④ $\frac{12}{\sqrt{6}} = \frac{12\times\sqrt{6}}{\sqrt{6}\times\sqrt{6}} = \frac{12\sqrt{6}}{6} = 2\sqrt{6}$ (since $\frac{12}{6} = 2$)

⑤ $\frac{50}{\sqrt{10}} = \frac{50\times\sqrt{10}}{\sqrt{10}\times\sqrt{10}} = \frac{50\sqrt{10}}{10} = 5\sqrt{10}$ (since $\frac{50}{10} = 50 \div 10 = 5$)

⑥ $\frac{7}{2\sqrt{14}} = \frac{7\times\sqrt{14}}{2\sqrt{14}\times\sqrt{14}} = \frac{7\times\sqrt{14}}{2\times14} = \frac{\sqrt{14}}{2\times2} = \frac{\sqrt{14}}{4}$ (since $\frac{7}{14} = \frac{7\div7}{14\div7} = \frac{1}{2}$)

Page 47

① $\frac{5\times7}{7\times15} = \frac{5}{15}\times\frac{7}{7} = \frac{1}{3}\times1 = \frac{1}{3}$

② $\frac{32\times6}{8\times3} = \frac{32}{8}\times\frac{6}{3} = 4\times2 = 8$

③ $\frac{5\times6}{24\times30} = \frac{5}{30}\times\frac{6}{24} = \frac{1}{6}\times\frac{6}{24} = \frac{6}{6}\times\frac{1}{24} = \frac{1}{24}$

④ $\frac{7\times12}{3\times28} = \frac{7}{28}\times\frac{12}{3} = \frac{1}{4}\times\frac{4}{1} = \frac{1}{1} = 1$

Page 48

① $\frac{1}{3}\times(6+9) = \frac{6+9}{3}$

② $\frac{3}{4}\times(28-16) = \frac{84-48}{4}$

③ $\frac{1}{2}\times\left(\frac{1}{3}+\frac{1}{5}\right) = \frac{1}{6}+\frac{1}{10}$

④ $\frac{3}{7}\times\left(\frac{4}{7}-\frac{2}{5}\right) = \frac{12}{49}-\frac{6}{35}$

Page 49

1. $\dfrac{7}{30} + \dfrac{11}{45} = \dfrac{7}{15 \times 2} + \dfrac{11}{15 \times 3} = \dfrac{1}{15} \times \left(\dfrac{7}{2} + \dfrac{11}{3}\right)$

2. $\dfrac{15}{16} - \dfrac{9}{20} = \dfrac{3 \times 5}{4 \times 4} - \dfrac{3 \times 3}{4 \times 5} = \dfrac{3}{4} \times \left(\dfrac{5}{4} - \dfrac{3}{5}\right)$

3. $\dfrac{8}{3} + \dfrac{12}{3} = \dfrac{4 \times 2}{3} + \dfrac{4 \times 3}{3} = \dfrac{4 \times (2 + 3)}{3}$

4. $\dfrac{6}{5} + \dfrac{2}{5} - \dfrac{1}{5} = \dfrac{6 + 2 - 1}{5}$

Page 50

1. $\dfrac{\frac{9}{16}}{\frac{3}{8}} = \dfrac{9}{16} \div \dfrac{3}{8} = \dfrac{9}{16} \times \dfrac{8}{3} = \dfrac{9 \times 8}{16 \times 3} = \dfrac{9}{3} \times \dfrac{8}{16} = \dfrac{3}{1} \times \dfrac{1}{2} = \dfrac{3}{2}$ (alternate answer: $1\frac{1}{2}$)

2. $\dfrac{8}{\frac{1}{4}} = 8 \div \dfrac{1}{4} = \dfrac{8}{1} \div \dfrac{1}{4} = \dfrac{8}{1} \times \dfrac{4}{1} = \dfrac{8 \times 4}{1 \times 1} = \dfrac{32}{1} = 32 \div 1 = 32$

3. $\dfrac{4\frac{1}{2}}{1\frac{3}{4}} = 4\dfrac{1}{2} \div 1\dfrac{3}{4} = \dfrac{9}{2} \div \dfrac{7}{4} = \dfrac{9}{2} \times \dfrac{4}{7} = \dfrac{9 \times 4}{2 \times 7} = \dfrac{36}{14} = \dfrac{36 \div 2}{14 \div 2} = \dfrac{18}{7}$ or $2\dfrac{4}{7}$

Page 51

1. $4 - \dfrac{6-3}{2+2} = \dfrac{4}{1} - \dfrac{3}{4} = \dfrac{4 \times 4}{1 \times 4} - \dfrac{3 \times 1}{4 \times 1} = \dfrac{16}{4} - \dfrac{3}{4} = \dfrac{16-3}{4} = \dfrac{13}{4}$ or $3\dfrac{1}{4}$

2. $6\dfrac{4}{5} - \left(-3\dfrac{2}{5}\right) = \dfrac{34}{5} - \left(-\dfrac{17}{5}\right) = \dfrac{34}{5} + \dfrac{17}{5} = \dfrac{34+17}{5} = \dfrac{51}{5}$ or $10\dfrac{1}{5}$

3. $3\dfrac{1}{6} \times \dfrac{10}{3} + \dfrac{2}{9} \div \left(\dfrac{1}{3}\right)^2 = \dfrac{19}{6} \times \dfrac{10}{3} + \dfrac{2}{9} \div \dfrac{1}{9} = \dfrac{190}{18} + \dfrac{2}{9} \times \dfrac{9}{1} = \dfrac{95}{9} + \dfrac{18}{9} = \dfrac{113}{9}$ or $12\dfrac{5}{9}$

4. $\left(-\dfrac{2}{3}\right)^2 - \left(-\dfrac{1}{2}\right)^3 = \dfrac{4}{9} - \left(-\dfrac{1}{8}\right) = \dfrac{4}{9} + \dfrac{1}{8} = \dfrac{4 \times 8}{9 \times 8} + \dfrac{1 \times 9}{8 \times 9} = \dfrac{32}{72} + \dfrac{9}{72} = \dfrac{32+9}{72} = \dfrac{41}{72}$

Note: $\left(-\dfrac{2}{3}\right)^2 = \left(-\dfrac{2}{3}\right) \times \left(-\dfrac{2}{3}\right) = \dfrac{4}{9}$ and $\left(-\dfrac{1}{2}\right)^3 = \left(-\dfrac{1}{2}\right) \times \left(-\dfrac{1}{2}\right) \times \left(-\dfrac{1}{2}\right) = -\dfrac{1}{8}$.

5. $\left(\dfrac{8}{3}\right)^{-1} + 4^{-2} = \dfrac{3}{8} + \dfrac{1}{4^2} = \dfrac{3}{8} + \dfrac{1}{16} = \dfrac{3 \times 2}{8 \times 2} + \dfrac{1}{16} = \dfrac{6}{16} + \dfrac{1}{16} = \dfrac{6+1}{16} = \dfrac{7}{16}$

6. $\dfrac{\frac{3}{4} + \frac{2}{3}}{\frac{1}{3} - \frac{1}{4}} = \dfrac{\frac{3 \times 3}{4 \times 3} + \frac{2 \times 4}{3 \times 4}}{\frac{1 \times 4}{3 \times 4} - \frac{1 \times 3}{4 \times 3}} = \dfrac{\frac{9+8}{12}}{\frac{4-3}{12}} = \dfrac{\frac{17}{12}}{\frac{1}{12}} = \dfrac{17}{12} \div \dfrac{1}{12} = \dfrac{17}{12} \times \dfrac{12}{1} = \dfrac{17}{1} \times \dfrac{12}{12} = 17 \times 1 = 17$

Page 52

1. $23 \div 3 = 7R2 = 7\dfrac{2}{3}$

2. $68 \div 7 = 9R5 = 9\dfrac{5}{7}$

3. $30 \div 4 = 7R2 = 7\dfrac{2}{4} = 7\dfrac{1}{2}$

4. $21 \div 9 = 2R3 = 2\dfrac{3}{9} = 2\dfrac{1}{3}$

5. $11 \div 2 = 5R1 = 5\dfrac{1}{2}$

Page 53

① (C) 18 since $6 \times 3 = 18$ and $9 \times 2 = 18$

② (B) $\frac{5}{32}$ since $\frac{3}{16} = \frac{6}{32}, \frac{1}{4} = \frac{8}{32}, \frac{7}{8} = \frac{28}{32}$, and $1 = \frac{32}{32}$

③ (C) $8\frac{4}{7}$ since $8 \times 7 = 56$ and $56 + 4 = 60$

④ (D) $\frac{29}{6}$ since $4 \times 6 + 5 = 24 + 5 = 29$

⑤ (D) $\frac{32}{8}$ since $\frac{21}{4} = \frac{21 \times 2}{4 \times 2} = \frac{42}{8}$ not $\frac{32}{8}$ (or since $\frac{32}{8} = 32 \div 8 = 4$)

⑥ (B) $3, \frac{17}{5}, 3\frac{1}{2}, \frac{11}{3}$ since $\frac{3 \times 30}{1 \times 30} = \frac{90}{30}, \frac{17 \times 6}{5 \times 6} = \frac{102}{30}, \frac{7 \times 15}{2 \times 15} = \frac{105}{30}$, and $\frac{11 \times 10}{3 \times 10} = \frac{110}{30}$

⑦ (C) $\frac{8}{19}$ since $2\frac{3}{8} = \frac{2 \times 8 + 3}{8} = \frac{19}{8}$

⑧ (C) $\frac{4}{5}$ pie since $4 \div 5 = \frac{4}{5}$

Page 54

⑨ (E) $\frac{53}{12}$ since $\frac{8}{3} + \frac{7}{4} = \frac{8 \times 4}{3 \times 4} + \frac{7 \times 3}{4 \times 3} = \frac{32}{12} + \frac{21}{12} = \frac{32 + 21}{12} = \frac{53}{12}$

⑩ (A) $\frac{2}{3}$ since $\frac{4}{5} - \frac{2}{15} = \frac{4 \times 3}{5 \times 3} - \frac{2 \times 1}{15 \times 1} = \frac{12}{15} - \frac{2}{15} = \frac{12 - 2}{15} = \frac{10}{15} = \frac{10 \div 5}{15 \div 5} = \frac{2}{3}$

⑪ (B) $\frac{8}{7}$ since $\frac{6}{7} \times \frac{4}{3} = \frac{6 \times 4}{7 \times 3} = \frac{24}{21} = \frac{24 \div 3}{21 \div 3} = \frac{8}{7}$

⑫ (A) $\frac{2}{3}$ since $\frac{4}{9} \div \frac{2}{3} = \frac{4}{9} \times \frac{3}{2} = \frac{4 \times 3}{9 \times 2} = \frac{12}{18} = \frac{12 \div 6}{18 \div 6} = \frac{2}{3}$

⑬ (E) $\frac{8}{125}$ since $\left(\frac{2}{5}\right)^3 = \frac{2 \times 2 \times 2}{5 \times 5 \times 5} = \frac{8}{125}$

⑭ (E) $\frac{16}{9}$ since $\left(\frac{3}{4}\right)^{-2} = \left(\frac{4}{3}\right)^2 = \frac{4 \times 4}{3 \times 3} = \frac{16}{9}$

⑮ (D) 36 pens since $\frac{3}{4} \times 48 = \frac{3}{4} \times \frac{48}{1} = \frac{3 \times 48}{4 \times 1} = \frac{144}{4} = \frac{144 \div 4}{4 \div 4} = \frac{36}{1} = 36$

Note: In word problems, the word "of" often means to multiply.

⑯ (E) $\frac{\sqrt{6}}{6}$ since $\frac{1}{\sqrt{6}} = \frac{1 \times \sqrt{6}}{\sqrt{6} \times \sqrt{6}} = \frac{\sqrt{6}}{6}$

Chapter 3: Decimals and Percents

Page 55

There aren't any problems on page 55.

Page 56

① The 8 in 2.1⬚8⬚7 is in the hundredths place.

② The 6 in ⬚6⬚25 is in the hundreds place.

③ The 9 in 0.⬚9⬚4 is in the tenths place.

④ The 1 in 0.00⬚1⬚23 is in the thousandths place.

⑤ The 7 in ⬚7⬚,214.638 is in the thousands place.

Page 57

① $0.2 > 0.08$ ② $88.3 < 88.7$ ③ $6.9 > 6.899$

④ $4.99 < 5.42$ ⑤ $7.0 = 7.00$ ⑥ $0.009 < 0.01$

⑦ $0.12, 0.194, 0.23, 0.3$

⑧ $2.6, 2.698, 2.703, 2.710$

Page 58

① $10^3 = 1000$ ② $10^{-3} = 0.001$ ③ $10^2 = 100$

④ $10^6 = 1,000,000$ ⑤ $10^0 = 1$ ⑥ $10^{-1} = 0.1$

⑦ $10^{-5} = 0.00001$ ⑧ $10^8 = 100,000,000$ ⑨ $10^{-4} = 0.0001$

⑩ $10^7 = 10,000,000$ ⑪ $10^{-6} = 0.000001$

⑫ $10^9 = 1,000,000,000$

Page 59

① 0.3   ② 3.8   ③ 0.6

④ 0.02   ⑤ 7.06   ⑥ 3.73

Page 60

$$\text{①} \; 2.3 + 3.8 \; \rightarrow \; \begin{array}{r} 2.3 \\ + 3.8 \\ \hline 6.1 \end{array}$$

$$\text{②} \; 4.5 - 1.75 \; \rightarrow \; \begin{array}{r} 4.50 \\ - 1.75 \\ \hline 2.75 \end{array}$$

$$\text{③} \; 0.7 + 0.04 \; \rightarrow \; \begin{array}{r} 0.70 \\ + 0.04 \\ \hline 0.74 \end{array}$$

$$\text{④} \; 0.06 - 0.005 \; \rightarrow \; \begin{array}{r} 0.060 \\ - 0.005 \\ \hline 0.055 \end{array}$$

$$\text{⑤} \; 9.86 + 0.7 \; \rightarrow \; \begin{array}{r} 9.86 \\ + 0.70 \\ \hline 10.56 \end{array}$$

$$\text{⑥} \; 1 - 0.003 \; \rightarrow \; \begin{array}{r} 1.000 \\ - 0.003 \\ \hline 0.997 \end{array}$$

Page 61

There aren't any problems on page 61.

Page 62

$$\text{①} \; 2 \text{ places} \quad \begin{array}{r} 73 \\ \times \; 6 \\ \hline 438 \end{array} \quad \begin{array}{r} 7.3 \\ \times 0.6 \\ \hline 4.38 \end{array}$$

$$\text{②} \; 2 \text{ places} \quad \begin{array}{r} 43 \\ \times 12 \\ \hline 86 \\ 430 \\ \hline 516 \end{array} \quad \begin{array}{r} 4.3 \\ \times 1.2 \\ \hline 5.16 \end{array}$$

$$\text{③} \; 4 \text{ places} \quad \begin{array}{r} 382 \\ \times \; 4 \\ \hline 1528 \end{array} \quad \begin{array}{r} 0.382 \\ \times \; 0.4 \\ \hline 0.1528 \end{array}$$

$$\text{④} \; 1 \text{ place} \quad \begin{array}{r} 32 \\ \times 3 \\ \hline 96 \end{array} \quad \begin{array}{r} 32 \\ \times 0.3 \\ \hline 9.6 \end{array}$$

$$\text{⑤} \; 3 \text{ places} \quad \begin{array}{r} 93 \\ \times \; 8 \\ \hline 744 \end{array} \quad \begin{array}{r} 0.93 \\ \times 0.8 \\ \hline 0.744 \end{array}$$

$$\text{⑥} \; 4 \text{ places} \quad \begin{array}{r} 27 \\ \times 19 \\ \hline 243 \\ 270 \\ \hline 513 \end{array} \quad \begin{array}{r} 0.27 \\ \times 0.19 \\ \hline 0.0513 \end{array}$$

$$\text{⑦} \; 3 \text{ places} \quad \begin{array}{r} 84 \\ \times 21 \\ \hline 84 \\ 1680 \\ \hline 1764 \end{array} \quad \begin{array}{r} 8.4 \\ \times 0.21 \\ \hline 1.764 \end{array}$$

$$\text{⑧} \; 2 \text{ places} \quad \begin{array}{r} 56 \\ \times 7 \\ \hline 392 \end{array} \quad \begin{array}{r} 0.56 \\ \times \; 7 \\ \hline 3.92 \end{array}$$

$$\text{⑨} \; 4 \text{ places} \quad \begin{array}{r} 15 \\ \times 6 \\ \hline 90 \end{array} \quad \begin{array}{r} 0.015 \\ \times 0.6 \\ \hline 0.0090 \end{array} \quad \begin{array}{r} 0.015 \\ \times 0.6 \\ \hline 0.009 \end{array}$$

$$\text{⑩} \; 5 \text{ places} \quad \begin{array}{r} 8 \\ \times 3 \\ \hline 24 \end{array} \quad \begin{array}{r} 0.008 \\ \times 0.03 \\ \hline 0.00024 \end{array}$$

Note: $0.0090 = 0.009$ since trailing zeroes don't matter in math (though they do matter for significant figures in science classes).

Page 63

$\text{\textcircled{1}}\ 0.74 \times 100\% = 74\%$ $\text{\textcircled{2}}\ 0.08 \times 100\% = 8\%$ $\text{\textcircled{3}}\ 1.9 \times 100\% = 190\%$

$\text{\textcircled{4}}\ 0.3 \times 100\% = 30\%$ $\text{\textcircled{5}}\ 4 \times 100\% = 400\%$ $\text{\textcircled{6}}\ 0.006 \times 100\% = 0.6\%$

$\text{\textcircled{7}}\ \dfrac{81\%}{100\%} = 0.81$ $\text{\textcircled{8}}\ \dfrac{230\%}{100\%} = 2.3$ $\text{\textcircled{9}}\ \dfrac{9\%}{100\%} = 0.09$

$\text{\textcircled{10}}\ \dfrac{70\%}{100\%} = 0.7$ $\text{\textcircled{11}}\ \dfrac{0.2\%}{100\%} = 0.002$ $\text{\textcircled{12}}\ \dfrac{5.6\%}{100\%} = 0.056$

Page 64

$\text{\textcircled{1}}\ 0.49 = \dfrac{49}{100}$ $\text{\textcircled{2}}\ 0.4 = \dfrac{4}{10} = \dfrac{4 \div 2}{10 \div 2} = \dfrac{2}{5}$

$\text{\textcircled{3}}\ 0.005 = \dfrac{5}{1000} = \dfrac{5 \div 5}{1000 \div 5} = \dfrac{1}{200}$ $\text{\textcircled{4}}\ 1.5 = \dfrac{15}{10} = \dfrac{15 \div 5}{10 \div 5} = \dfrac{3}{2}$

$\text{\textcircled{5}}\ 0.75 = \dfrac{75}{100} = \dfrac{75 \div 25}{100 \div 25} = \dfrac{3}{4}$ $\text{\textcircled{6}}\ 0.024 = \dfrac{24}{1000} = \dfrac{24 \div 8}{1000 \div 8} = \dfrac{3}{125}$

$\text{\textcircled{7}}\ 2.88 = \dfrac{288}{100} = \dfrac{288 \div 4}{100 \div 4} = \dfrac{72}{25}$ $\text{\textcircled{8}}\ 0.96 = \dfrac{96}{100} = \dfrac{96 \div 4}{100 \div 4} = \dfrac{24}{25}$

Page 65

$\text{\textcircled{1}}\ \dfrac{3}{4} = \dfrac{3 \times 25}{4 \times 25} = \dfrac{75}{100} = 0.75$ $\text{\textcircled{2}}\ \dfrac{27}{50} = \dfrac{27 \times 2}{50 \times 2} = \dfrac{54}{100} = 0.54$

$\text{\textcircled{3}}\ \dfrac{7}{2} = \dfrac{7 \times 5}{2 \times 5} = \dfrac{35}{10} = 3.5$ $\text{\textcircled{4}}\ \dfrac{16}{125} = \dfrac{16 \times 8}{125 \times 8} = \dfrac{128}{1000} = 0.128$

$\text{\textcircled{5}}\ \dfrac{31}{25} = \dfrac{31 \times 4}{25 \times 4} = \dfrac{124}{100} = 1.24$ $\text{\textcircled{6}}\ \dfrac{11}{5} = \dfrac{11 \times 2}{5 \times 2} = \dfrac{22}{10} = 2.2$

$\text{\textcircled{7}}\ \dfrac{17}{40} = \dfrac{17 \times 25}{40 \times 25} = \dfrac{425}{1000} = 0.425$ $\text{\textcircled{8}}\ \dfrac{543}{200} = \dfrac{543 \times 5}{200 \times 5} = \dfrac{2715}{1000} = 2.715$

Page 66

$\text{\textcircled{1}}\ 0.\overline{24} = \dfrac{24}{99} = \dfrac{24 \div 3}{99 \div 3} = \dfrac{8}{33}$ $\text{\textcircled{2}}\ 0.\overline{6} = \dfrac{6}{9} = \dfrac{6 \div 3}{9 \div 3} = \dfrac{2}{3}$

$\text{\textcircled{3}}\ 0.\overline{05} = \dfrac{5}{99}$ $\text{\textcircled{4}}\ 0.0\overline{5} = \dfrac{1}{10}\left(0.\overline{5}\right) = \dfrac{1}{10}\left(\dfrac{5}{9}\right) = \dfrac{5}{90} = \dfrac{5 \div 5}{90 \div 5} = \dfrac{1}{18}$

Note: $0.\overline{05}$ means 0.05050505…, whereas $0.0\overline{5}$ means 0.05555555…

$\text{\textcircled{5}}\ 0.\overline{110} = \dfrac{110}{999}$ $\text{\textcircled{6}}\ 1.\overline{3} = 1 + 0.\overline{3} = 1 + \dfrac{3}{9} = 1 + \dfrac{3 \div 3}{9 \div 3} = 1\dfrac{1}{3} = \dfrac{4}{3}$

Page 67

$\text{\textcircled{1}}\ \dfrac{25}{33} = \dfrac{25 \times 3}{33 \times 3} = \dfrac{75}{99} = 0.\overline{75}$ $\text{\textcircled{2}}\ \dfrac{14}{111} = \dfrac{14 \times 9}{111 \times 9} = \dfrac{126}{999} = 0.\overline{126}$

$\text{\textcircled{3}}\ \dfrac{1}{30} = \dfrac{1 \times 3}{30 \times 3} = \dfrac{3}{90} = \dfrac{1}{10}\left(\dfrac{3}{9}\right) = \dfrac{1}{10}\left(0.\overline{3}\right) = 0.0\overline{3}$ (meaning 0.03333333…)

$\text{\textcircled{4}}\ \dfrac{14}{9} = 1\dfrac{5}{9} = 1 + \dfrac{5}{9} = 1 + 0.\overline{5} = 1.\overline{5}$

⑤ $\frac{12}{11} = 1\frac{1}{11} = 1 + \frac{1}{11} = 1 + \frac{1\times9}{11\times9} = 1 + \frac{9}{99} = 1 + 0.\overline{09} = 1.\overline{09}$

Notes: $\frac{9}{99}$ must have two digits repeating because 99 has two 9's. Since 9 is just one digit, you must think of 9 as 09 to get the second digit. $1.\overline{09}$ means 1.0909090909...

⑥ $\frac{1}{6} = \frac{1\times15}{6\times15} = \frac{15}{90} = \frac{1}{10}\frac{15}{9} = \frac{1}{10}\left(1\frac{6}{9}\right) = \frac{1}{10}\left(1 + \frac{6}{9}\right) = \frac{1}{10}\left(1 + 0.\overline{6}\right) = \frac{1}{10}1.\overline{6} = 0.1\overline{6}$

Page 68

① $\frac{0.3}{2.5} = \frac{0.3\times10}{2.5\times10} = \frac{3}{25} = \frac{3\times4}{25\times4} = \frac{12}{100} = 0.12$

② $\frac{0.32}{0.4} = \frac{0.32\times100}{0.4\times100} = \frac{32}{40} = \frac{32\div8}{40\div8} = \frac{4}{5} = \frac{4\times2}{5\times2} = \frac{8}{10} = 0.8$

③ $\frac{1}{0.05} = \frac{1\times100}{0.05\times100} = \frac{100}{5} = \frac{100\div5}{5\div5} = \frac{20}{1} = 20$

④ $\frac{2.48}{0.8} = \frac{2.48\times100}{0.8\times100} = \frac{248}{80} = \frac{248\div8}{80\div8} = \frac{31}{10} = 3.1$

Page 69

① $1.4 < \frac{3}{2}$ since $\frac{3}{2} = \frac{3\times5}{2\times5} = \frac{15}{10} = 1.5$ and $1.4 < 1.5$

② $0.5\% < 0.02$ since $0.5\% = \frac{0.5\%}{100\%} = 0.005 < 0.02$

③ $\frac{7}{20} > 30\%$ since $\frac{7}{20} = \frac{7\times5}{20\times5} = \frac{35}{100} = 0.35 = 0.35\times100\% = 35\% > 30\%$

④ $35\%, 0.53, \frac{3}{5}$ since $\frac{3}{5} = \frac{3\times2}{5\times2} = \frac{6}{10} = 0.6$ and $35\% = \frac{35\%}{100\%} = 0.35$

⑤ $120\%, \frac{5}{2}, 2.7$ since $\frac{5}{2} = \frac{5\times5}{2\times5} = \frac{25}{10} = 2.5$ and $120\% = \frac{120\%}{100\%} = 1.2$

Page 70

① 27.2 is rational

② 4,315,986 is both whole and rational

③ $\frac{1}{3}$ is rational

④ $\sqrt{3}$ is irrational

⑤ $\frac{12}{4} = 12 \div 4 = 3$ is both whole and rational

⑥ $\sqrt{16} = 4$ (since $4^2 = 4 \times 4 = 16$; see Sec. 1.16) is both whole and rational

Page 71

① $3.4 + (-0.72) - (-1.45) = 3.4 - 0.72 + 1.45 = \boxed{4.13}$

② $12 \times 0.3 - \frac{1}{2} \times 10 \times 0.3^2 = 12 \times 0.3 - \frac{1}{2} \times 10 \times 0.09 = 3.6 - 5 \times 0.09$

$= 3.6 - 0.45 = \boxed{3.15}$

③ $8.1 - (1.75 + 0.2 \div 0.8)^2 = 8.1 - (1.75 + 0.25)^2 = 8.1 - 2^2 = 8.1 - 4$

$= \boxed{4.1}$

④ $(0.4)^{-1} + (0.2)^{-1} = \left(\frac{4}{10}\right)^{-1} + \left(\frac{2}{10}\right)^{-1} = \frac{10}{4} + \frac{10}{2} = \frac{10\times25}{4\times25} + \frac{10\times5}{2\times5} = \frac{250}{100} + \frac{50}{10}$

$= 2.5 + 5 = \boxed{7.5}$

⑤ $1.6 \times \frac{7-3.8}{0.65+0.15} = 1.6 \times \frac{3.2}{0.8} = 1.6 \times 4 = \boxed{6.4}$

⑥ $10^{-6} \times \frac{10^4 \times 10^5}{(10^3)^2} = \frac{1}{10^6} \times \frac{10^4 \times 10^5}{10^3 \times 10^3} = \frac{10,000 \times 100,000}{1,000,000 \times 1,000 \times 1,000} = \frac{1}{1,000} = \boxed{0.001}$

Page 72

① (C) hundredths

② (B) $0.005, 0.008, 0.079, 0.097, 0.6$

③ (B) 0.001 since $10^{-3} = \frac{1}{1000} = 0.001$

④ (D) 0.046 since the 5 is in the thousandths place and it rounds up to 6 because 7 is 5 or more

⑤ (B) 0.356 since $0.34 + 0.016$ $\rightarrow$
$$\begin{array}{r} 0.340 \\ +\ 0.016 \\ \hline 0.356 \end{array}$$

⑥ (D) 2.86 since $3.1 - 0.24$ $\rightarrow$
$$\begin{array}{r} 3.10 \\ -\ 0.24 \\ \hline 2.86 \end{array}$$

⑦ (B) 0.222 since $\begin{array}{r} 7.4 \\ \times\ 0.03 \\ \hline \end{array}$ $\xrightarrow{\text{3 places}}$ $\begin{array}{r} 74 \\ \times\ 3 \\ \hline 222 \end{array}$ $\begin{array}{r} 7.4 \\ \times\ 0.03 \\ \hline 0.222 \end{array}$

⑧ (E) 2.4 since $0.48 \div 0.2 = \frac{0.48}{0.2} = \frac{0.48\times100}{0.2\times100} = \frac{48}{20} = \frac{48\div2}{20\div2} = \frac{24}{10} = 2.4$

Page 73

⑨ (D) 37.5% since $0.375 = 0.375 \times 100\% = 37.5\%$

⑩ (B) 1.4 since $140\% = \frac{140\%}{100\%} = 1.4$

⑪ (D) 0.6 since $\frac{3}{5} = \frac{3\times2}{5\times2} = \frac{6}{10} = 0.6$

⑫ (D) $\frac{34}{25}$ since $1.36 = \frac{136}{100} = \frac{136\div4}{100\div4} = \frac{34}{25}$

⑬ (B) 11.8% since $\frac{59}{50} = 1\frac{9}{50} = \frac{59\times2}{50\times2} = \frac{118}{100} = 1.18 = 1.18 \times 100\% = 118\%$

⑭ (E) $\sqrt{10}$

⑮ (A) $3, 337\%, \frac{27}{8}, 3.38, 3\frac{1}{2}$ since $337\% = \frac{337\%}{100\%} = 3.37, 3\frac{1}{2} = \frac{7}{2} = \frac{7\times5}{2\times5} = \frac{35}{10} = 3.5$

and $\frac{27}{8} = \frac{27\times125}{8\times125} = \frac{3375}{1000} = 3.375$

⑯ (B) 0.256 since $32\% \times \frac{4}{5} = \frac{32\%}{100\%} \times \frac{4\times2}{5\times2} = 0.32 \times \frac{8}{10} = 0.32 \times 0.8$

and $\begin{array}{r} 0.32 \\ \times\, 0.8 \\ \hline \end{array}$ $\xrightarrow{\text{3 places}}$ $\begin{array}{r} 32 \\ \times\, 8 \\ \hline 256 \end{array}$ $\begin{array}{r} 0.32 \\ \times\, 0.8 \\ \hline 0.256 \end{array}$

Note: In word problems, the word "of" often means to multiply.

Page 74

⑰ (D) a rational number (like $\frac{2}{3}$ or $\frac{7}{4}$)

⑱ (A) 0.4 MB since $6.4 \div 16 = \frac{6.4}{16} = \frac{6.4\times10}{16\times10} = \frac{64}{160} = \frac{64\div16}{160\div16} = \frac{4}{10} = 0.4$

Note: MB stands for megabytes, but it doesn't really matter what MB is.

⑲ (C) 25% There are two parts to this answer.

First, $\frac{3}{4} = \frac{3\times25}{4\times25} = \frac{75}{100} = 0.75 = 0.75 \times 100\% = 75\%$, but this isn't the answer.

Subtract 75% from 100% to find the number who **DON'T** take the bus:

$100\% - 75\% = 25\%$ (It's a logic problem: If 75% take the bus, 25% don't.)

⑳ (E) 78% since $\frac{39}{50} = \frac{39\times2}{50\times2} = \frac{78}{100} = 0.78 = 0.78 \times 100\% = 78\%$

Chapter 4: Proportions

Page 75

There aren't any problems on page 75.

Page 76

(1) $2:3 = \frac{2}{3}$ (2) $1:5 = \frac{1}{5}$ (3) $6:11 = \frac{6}{11}$

(4) $4:1 = \frac{4}{1} = 4$ (5) $3:8 = \frac{3}{8}$ (6) $14:5 = \frac{14}{5}$

(7) $\frac{3}{8} = 3:8$ (8) $\frac{1}{2} = 1:2$ (9) $\frac{9}{4} = 9:4$

(10) $\frac{1}{4} = 1:4$ (11) $\frac{3}{2} = 3:2$ (12) $\frac{11}{16} = 11:16$

Page 77

(1) $4:12 = \frac{4}{12} = \frac{4 \div 4}{12 \div 4} = \frac{1}{3} = 1:3$

(2) $15:10 = \frac{15}{10} = \frac{15 \div 5}{10 \div 5} = \frac{3}{2} = 3:2$

(3) $18:24 = \frac{18}{24} = \frac{18 \div 6}{24 \div 6} = \frac{3}{4} = 3:4$

(4) $35:21 = \frac{35}{21} = \frac{35 \div 7}{21 \div 7} = \frac{5}{3} = 5:3$

(5) $30:6 = \frac{30}{6} = \frac{30 \div 6}{6 \div 6} = \frac{5}{1} = 5:1$

Page 78

(1)

apples	5	10	15	20	25
oranges	2	4	6	8	10

(2)

wins	9	18	27	36	45
losses	5	10	15	20	25

Page 79

There aren't any problems on page 79.

Page 80

① $\frac{60 \text{ girls}}{4 \text{ girls}} = 60 \div 4 = 15$ (**NOT** the final answer)

Multiply 4 girls by 15 to make 60 girls.

$4{:}3 = \frac{4}{3} = \frac{4 \times 15}{3 \times 15} = \frac{60}{45} = 60{:}45 \rightarrow \boxed{45 \text{ boys}}$

② $\frac{960 \text{ students}}{240 \text{ students}} = 960 \div 240 = 4$ (**NOT** the final answer)

Divide 960 students by 4 to make 240 students.

$12{:}960 = \frac{12}{960} = \frac{12 \div 4}{960 \div 4} = \frac{3}{240} = 3{:}240 \rightarrow \boxed{3 \text{ bus drivers}}$

③ $\frac{320}{48} = \frac{320 \div 16}{48 \div 16} = \frac{20}{3} = \boxed{20{:}3}$ The ratio is 20 cows to 3 horses (20:3).

Page 81

There aren't any problems on page 81.

Page 82

① 70 children + 28 adults = 98 family members

$\frac{70 \text{ children}}{98 \text{ family members}} = \frac{70}{98} = \frac{70 \div 14}{98 \div 14} = \frac{5}{7} = \boxed{5{:}7}$

② $\frac{140 \text{ writing utensils}}{7 \text{ writing utensils}} = 140 \div 7 = 20$ (**NOT** the final answer)

Multiply 7 writing utensils by 20 to make 140 writing utensils.

$4{:}7 = \frac{4}{7} = \frac{4 \times 20}{7 \times 20} = \frac{80}{140} = 80{:}140 \rightarrow \boxed{80 \text{ pens}}$

③ $\frac{120 \text{ desktops}}{3 \text{ desktops}} = 120 \div 3 = 40$ (**NOT** the final answer)

Multiply 3 desktops by 40 to make 120 desktops.

$3{:}4 = \frac{3}{4} = \frac{3 \times 40}{4 \times 40} = \frac{120}{160} = 120{:}160 \rightarrow \boxed{160 \text{ computers}}$

Page 83

① 800 miles in 9 hours $= \frac{800 \text{ miles}}{9 \text{ hours}}$

② 20 mintues for 3 problems $= \frac{20 \text{ minutes}}{3 \text{ problems}}$

③ 24 seeds in 1 package $= \frac{24 \text{ seeds}}{1 \text{ package}} = 24 \frac{\text{seeds}}{\text{package}}$ since $\frac{24}{1} = 24$

④ 5 dollars for 12 eggs $= \frac{5 \text{ dollars}}{12 \text{ eggs}}$

Page 84

①

dollars	125	250	375	500	625
hours	8	16	24	32	40

②

beats	5	10	15	20	25
seconds	4	8	12	16	20

Page 85

① $\dfrac{3 \text{ inches}}{50 \text{ days}} = \dfrac{3 \times 2 \text{ inches}}{50 \times 2 \text{ days}} = \dfrac{6 \text{ inches}}{100 \text{ days}} = 0.06$ inches per day

② $\dfrac{210 \text{ miles}}{4 \text{ hours}} = \dfrac{210 \times 25 \text{ miles}}{4 \times 25 \text{ hours}} = \dfrac{5250 \text{ miles}}{100 \text{ hours}} = 52.5$ miles per hour

Note: We may ignore the trailing zero in 52.50.

③ $\dfrac{94 \text{ dollars}}{10 \text{ hours}} = 9.4$ dollars per hour

④ $\dfrac{7 \text{ loaves}}{20 \text{ minutes}} = \dfrac{7 \times 5 \text{ loaves}}{20 \times 5 \text{ minutes}} = \dfrac{35 \text{ loaves}}{100 \text{ minutes}} = 0.35$ loaves per minute

Page 86

There aren't any problems on page 86.

Page 87

① $r = \dfrac{d}{t} = \dfrac{280 \text{ miles}}{7 \text{ hours}} = 40$ miles per hour

② $d = r \times t = 18 \dfrac{\text{kilometers}}{\text{hour}} \times 1.5 \text{ hours} = 27$ kilometers

③ $t = \dfrac{d}{r} = \dfrac{200 \text{ yards}}{8 \text{ yards/second}} = 25$ seconds

Page 88

There aren't any problems on page 88.

Page 89

① Since Q is over V, make a pyramid with Q at the top.

	Q	
C		V

When you cover Q, since C is beside V you get $Q = C \times V$.

② Since I multiplies R, make a pyramid with I beside R.

	V	
I		R

When you cover I, since V is over R you get $I = \dfrac{V}{R}$.

③ Since c is over v, make a pyramid with c at the top.

	c	
n		v

When you cover v, since c is over n you get $v = \dfrac{c}{n}$.

Page 90

There aren't any problems on page 90.

Page 91

① Since I is over O, make a pyramid with I at the top.

	I	
M		O

When you cover I, since M is beside O you get $I = M \times O$.

$I = M \times O = 8 \times 40 \text{ cm} = 320 \text{ cm}$

② Since $rise$ is over run, make a pyramid with $rise$ at the top.

	$rise$	
m		run

When you cover run, since $rise$ is over m you get $run = \dfrac{rise}{m}$.

$run = \dfrac{rise}{m} = \dfrac{-12}{4} = -3$

Page 92

There aren't any problems on page 92.

Page 93

① $80 \dfrac{\text{words}}{\text{min.}} = \dfrac{80 \text{ words}}{1 \text{ min.}} = \dfrac{80 \text{ words} \times 7}{1 \text{ min.} \times 7} = \dfrac{560 \text{ words}}{7 \text{ min.}} \rightarrow 560$ words

② $3.25 \dfrac{\text{rev.}}{\text{sec.}} = \dfrac{325 \text{ rev.}}{100 \text{ sec.}} = \dfrac{325 \text{ rev.} \div 25}{100 \text{ sec.} \div 25} = \dfrac{13 \text{ rev.}}{4 \text{ sec.}} = \dfrac{13 \text{ rev.} \times 6}{4 \text{ sec.} \times 6} = \dfrac{78 \text{ rev.}}{24 \text{ sec.}} \rightarrow 78$ times

③ $42.5 \dfrac{\text{mi.}}{\text{gal.}} = \dfrac{425 \text{ mi.}}{10 \text{ gal.}} = \dfrac{425 \text{ mi.} \div 5}{10 \text{ gal.} \div 5} = \dfrac{85 \text{ mi.}}{2 \text{ gal.}} = \dfrac{85 \text{ mi.} \times 4}{2 \text{ gal.} \times 4} = \dfrac{340 \text{ mi.}}{8 \text{ gal.}} \rightarrow 340$ miles

Page 94

There aren't any problems on page 94.

Page 95

① Since $\frac{4}{3} = \frac{4\times8}{3\times8} = \frac{32}{24}$, it follows that $\boxed{} = 32$.

② Since $\frac{20}{32} = \frac{20\div4}{32\div4} = \frac{5}{8}$, it follows that $\boxed{} = 5$.

③ Since $\frac{7}{11} = \frac{7\times3}{11\times3} = \frac{21}{33}$, it follows that $\boxed{} = 33$.

④ Since $\frac{220}{100} = \frac{220\div20}{100\div20} = \frac{11}{5}$, it follows that $\boxed{} = 5$.

⑤ Since $\frac{75}{45} = \frac{75\div15}{45\div15} = \frac{5}{3} = \frac{5\times12}{3\times12} = \frac{60}{36}$, it follows that $\boxed{} = 60$.

Page 96

There aren't any problems on page 96.

Page 97

① Since $\frac{100\ \text{beads}}{3\ \text{jars}} = \frac{100\ \text{beads}\times6}{3\ \text{jars}\times6} = \frac{600\ \text{beads}}{18\ \text{jars}}$, they could hold 600 beads.

② Since $\frac{150\ \text{days}}{7\ \text{vac.}} = \frac{150\ \text{work}\times3}{7\ \text{vac.}\times3} = \frac{450\ \text{work}}{21\ \text{vac.}}$, she would need to work 450 days.

③ Since $\frac{\$429}{200\ \text{pages}} = \frac{\$429\times4}{200\ \text{pages}\times4} = \frac{\$1716}{800\ \text{pages}}$, it would cost $1716.

Page 98

① (B) 4:3 since $24:18 = \frac{24}{18} = \frac{24\div6}{18\div6} = \frac{4}{3} = 4:3$

② (A) 2:3 since $9:15 = \frac{9}{15} = \frac{9\div3}{15\div3} = \frac{3}{5} = \frac{3\times2}{5\times2} = \frac{6}{10} = \frac{6\times2}{10\times2} = \frac{12}{20} = \frac{3\times7}{5\times7} = \frac{21}{35}$

③ (C) 12 since $3:2 = \frac{3\ \text{dresses}}{2\ \text{shirts}} = \frac{3\times6}{2\times6} = \frac{18\ \text{dresses}}{12\ \text{shirts}}$

④ (D) 40 since $5:2 = \frac{5\ \text{apples}}{2\ \text{cups}} = \frac{5\times8}{2\times8} = \frac{40\ \text{apples}}{16\ \text{cups}}$

⑤ (B) 5:13 since 40 dogs + 25 cats = 65 pets and since

25 cats:65 pets $= \frac{25\ \text{cats}}{65\ \text{pets}} = \frac{25\div5}{65\div5} = \frac{5\ \text{cats}}{13\ \text{pets}}$

⑥ (A) 35 since $7:10 = \frac{7}{10} = \frac{7\times5}{10\times5} = \frac{35}{50} = 35:50$

⑦ (A) $\frac{2\ \text{in.}}{3\ \text{mo.}}$ since $\frac{8\ \text{in.}}{12\ \text{mo.}} = \frac{8\div4}{12\div4} = \frac{2\ \text{in.}}{3\ \text{mo.}}$

Page 99

⑧ (D) 4.32 since $\dfrac{108 \text{ words}}{25 \text{ sec.}} = \dfrac{108 \times 4}{25 \times 4} = \dfrac{432 \text{ words}}{100 \text{ sec.}} = 4.32$ words per sec

⑨ (E) 480 mi. since $d = r \times t = 60\,\dfrac{\text{mi.}}{\text{hr}} \times 8 \text{ hr} = 480 \text{ mi.}$

⑩ (D) 80 s since $t = \dfrac{d}{r} = \dfrac{240 \text{ m}}{3 \text{ m/s}} = 80$ s

⑪ (B) 8 in./hr since $r = \dfrac{d}{t} = \dfrac{32 \text{ in.}}{4 \text{ hr}} = 8$ in. per hr

⑫ (D) \$49.50 since \$8.25 per mo. $= \dfrac{\$8.25}{1 \text{ mo.}} = \dfrac{\$8.25 \times 6}{1 \text{ mo.} \times 6} = \dfrac{\$49.50}{6 \text{ mo.}}$

Note: $\$8.25 \times 6 = (\$8 + \$0.25) \times 6 = \$8 \times 6 + \$0.25 \times 6 = \$48 + \$1.50 = \$49.50$

⑬ (B) 21 since $\dfrac{7 \text{ sketches}}{30 \text{ min.}} = \dfrac{7 \times 3}{30 \times 3} = \dfrac{21 \text{ sketches}}{90 \text{ min.}}$

⑭ (C) 36 since $\dfrac{3}{4} = \dfrac{3 \times 12}{4 \times 12} = \dfrac{36}{48}$

⑮ (C) 18 since $\dfrac{24}{72} = \dfrac{24 \div 4}{72 \div 4} = \dfrac{6}{18}$

Page 100

⑯ (C) 18 mi. since $\dfrac{3 \text{ in.}}{12 \text{ mi.}} = \dfrac{3 \times 1.5}{12 \times 1.5} = \dfrac{4.5 \text{ in.}}{18 \text{ mi.}}$

⑰ (B) 15 since $\dfrac{\text{buy } 3}{\text{get 2 free}} = \dfrac{3 \times 5}{2 \times 5} = \dfrac{\text{buy } 15}{\text{get 10 free}}$

⑱ (B) 45 since $\dfrac{27 \text{ lines}}{36 \text{ min.}} = \dfrac{27 \div 9}{36 \div 9} = \dfrac{3 \text{ lines}}{4 \text{ min.}} = \dfrac{3 \times 15}{4 \times 15} = \dfrac{45 \text{ lines}}{60 \text{ min.}}$

⑲ (D) 7 hr since $\dfrac{125 \text{ nwsp}}{5 \text{ hr}} = \dfrac{125 \div 5}{5 \div 5} = \dfrac{25 \text{ nwsp}}{1 \text{ hr}} = \dfrac{25 \times 7}{1 \times 7} = \dfrac{175 \text{ nwsp}}{7 \text{ hr}}$

⑳ (E) 225 mi. since $\dfrac{180 \text{ mi.}}{4 \text{ hr}} = \dfrac{125 \div 4}{5 \div 4} = \dfrac{45 \text{ mi.}}{1 \text{ hr}} = \dfrac{45 \times 5}{1 \times 5} = \dfrac{225 \text{ mi.}}{5 \text{ hr}}$

Note: $45 \times 5 = (40 + 5) \times 5 = 40 \times 5 + 5 \times 5 = 200 + 25 = 225$

Chapter 5: Variables

Page 101

There aren't any problems on page 101.

Page 102

① t is the variable.

② 3, 2, and 8 are all constants.

③ -5 is the coefficient.

④ y and z are the variables.

⑤ 8 and 1 are the coefficients (since $1y = y$, since 1 times y equals y).

Page 103

① $3x^2 + 2x - 1$ means "3 times x squared plus 2 times x minus 1."

② $-x + \frac{9}{x}$ means "negative x plus 9 divided by x."

③ $x(x + 6)$ means "x times quantity x plus 6."

④ $\frac{x^2 - 4}{2x + 3}$ means "quantity x squared minus 4 divided by quantity 2 times x plus 3."

⑤ $(2x + 1)^2(3x)$ means "2 times x plus 1 quantity squared divided by quantity 3 times x."

Note: An alternative to using the word "quantity" is to say "open parentheses" and "closed parentheses."

Page 104

① $x - 1 = 1$ is an equation.

② $(x + 3)(2x + 1)$ is an expression.

③ $x^2 - 3x + 2$ is an expression.

④ $(3x - 2)^2 = 100$ is an equation.

⑤ $\frac{x}{2} = \frac{7}{6}$ is an equation.

⑥ $\frac{4x - 3}{x} - \frac{x}{2}$ is an expression.

Page 105

① $x - 5$ ② $2x$ ③ $4x$

④ $\sqrt{x}$ ⑤ $\dfrac{x}{y^2}$ or $\dfrac{d}{t^2}$ ⑥ xy

Page 106

① $S = 6L^2 = 6(3)^2 = 6(9) = 54$

② $v = \dfrac{d}{t} = \dfrac{42}{7} = 6$

③ $P = b^c = 10^4 = (10)(10)(10)(10) = 10,000$

④ $a = \dfrac{1}{2}gt^2 = \dfrac{1}{2}(10)(4)^2 = \dfrac{1}{2}(10)(16) = \dfrac{160}{2} = 80$

⑤ $I = \dfrac{V}{R} = \dfrac{12}{6} = 2$

⑥ $w = \dfrac{xy}{z} = \dfrac{(2)(6)}{4} = \dfrac{12}{4} = 3$

Page 107

There aren't any problems on page 107.

Page 108

① $x - 6 + 6 = 3 + 6$

$x = \boxed{9}$

Check: $x - 6 = 9 - 6 = 3$

③ $\dfrac{x}{4}4 = 9(4)$

$x = \boxed{36}$

Check: $\dfrac{x}{4} = \dfrac{36}{4} = 9$

⑤ $x + 9 - 9 = 14 - 9$

$x = \boxed{5}$

Check: $x + 9 = 5 + 9 = 14$

⑦ $\dfrac{7x}{7} = \dfrac{49}{7}$

$x = \boxed{7}$

Check: $7x = 7(7) = 49$

② $\dfrac{2x}{2} = \dfrac{10}{2}$

$x = \boxed{5}$

Check: $2(5) = 10$

④ $x + 2 - 2 = 5 - 2$

$x = \boxed{3}$

Check: $x + 2 = 3 + 2 = 5$

⑥ $x - 4 + 4 = 8 + 4$

$x = \boxed{12}$

Check: $x - 4 = 12 - 4 = 8$

⑧ $\dfrac{x}{8}8 = 7(8)$

$x = \boxed{56}$

Check: $\dfrac{56}{8} = 7$

Page 109

  ① $(8 - 5)x + 3 + 6 = 3x + 9$

  ② $(2 + 1)x^2 + (4 + 3)x = 3x^2 + 7x$

  ③ $(9 - 7)x^2 - 9 - 7 = 2x^2 - 16$

  ④ $(1 + 4)x^2 + (2 - 2)x + 3 - 1 = 5x^2 + 0x + 2 = 5x^2 + 2$

Page 110

There aren't any problems on page 110.

Page 111

① $9x + 1 = 7x + 7$

Subtract 1 and $7x$.

$9x - 7x = 7 - 1$

$2x = 6$

Divide by 2.

$x = \frac{6}{2} = \boxed{3}$

Check: $9(3) + 1 = 28 = 7(3) + 7$

② $3x - 8 = 2x - 4$

Add 8 and subtract $2x$.

$3x - 2x = -4 + 8$

$x = \boxed{4}$

Check: $3(4) - 8 = 4 = 2(4) - 4$

③ $5x - 9 = 3x + 9$

Add 9 and subtract $3x$.

$5x - 3x = 9 + 9$

$2x = 18$

Divide by 2.

$x = \frac{18}{2} = \boxed{9}$

Check: $5(9) - 9 = 36 = 3(9) + 9$

④ $x + 3 = 2x - 5$

Add 5 and subtract x.

$3 + 5 = 2x - x$

$\boxed{8} = x$

Check: $8 + 3 = 11 = 2(8) - 5$

⑤ $8x - 24 = 7x - 3x$

Add 24 and $3x$. Subtract $7x$.

$8x + 3x - 7x = 0 + 24$

$4x = 24$

Divide by 4.

$x = \frac{24}{4} = \boxed{6}$

Check: $8(6) - 24 = 24 = 7(6) - 3(6)$

⑥ $2x - 6 = 12$

Add 6.

$2x = 12 + 6 = 18$

Divide by 2.

$x = \frac{18}{2} = \boxed{9}$

Check: $2(9) - 6 = 12$

⑦ $6x - 3 + 4x = 7x + 9$

Add 3 and subtract $7x$.

$6x + 4x - 7x = 9 + 3$

$3x = 12$

Divide by 3.

$x = \frac{12}{3} = \boxed{4}$

Check: $6(4) - 3 + 4(4) = 37 = 7(4) + 9$

⑧ $50 - 8x = 15 - 3x$

Add $8x$ and subtract 15.

$50 - 15 = -3x + 8x$

$35 = 5x$

Divide by 5.

$\frac{35}{5} = \boxed{7} = x$

Check: $50 - 8(7) = -6 = 15 - 3(7)$

Page 112

There aren't any problems on page 112.

Page 113

① $8x + 2 = 6x + 9$

Subtract 2 and $6x$.

$8x - 6x = 9 - 2$

$2x = 7$

Divide by 2.

$x = \boxed{\frac{7}{2}} = \boxed{3\frac{1}{2}}$

Check: $8\left(\frac{7}{2}\right) + 2 = 30 = 6\left(\frac{7}{2}\right) + 9$

② $x - \frac{1}{3} = \frac{1}{6}$

Add $\frac{1}{3}$.

$x = \frac{1}{6} + \frac{1}{3} = \frac{1}{6} + \frac{1}{3}\left(\frac{2}{2}\right) = \frac{1}{6} + \frac{2}{6} = \frac{3}{6} = \boxed{\frac{1}{2}}$

Check: $\frac{1}{2} - \frac{1}{3} = \frac{3}{6} - \frac{2}{6} = \frac{1}{6}$

③ $2x + 8 = 0$

Subtract 8.

$2x = 0 - 8$

$2x = -8$

Divide by 2.

$x = -\frac{8}{2} = \boxed{-4}$

Check: $2(-4) + 8 = 0$

④ $3x - 4 = 7x + 6$

Subtract $3x$ and 6.

$-4 - 6 = 7x - 3x$

$-10 = 4x$

Divide by 4.

$-\frac{10}{4} = \boxed{-\frac{5}{2}} = \boxed{-2\frac{1}{2}} = x$

Check: $3\left(-\frac{5}{2}\right) - 4 = -\frac{23}{2} = 7\left(-\frac{5}{2}\right) + 6$

⑤ $3x + \frac{1}{2} = \frac{1}{4} + 5x$

Subtract $\frac{1}{4}$ and $3x$.

$\frac{1}{2} - \frac{1}{4} = 5x - 3x$

$\frac{1}{2}\left(\frac{2}{2}\right) - \frac{1}{4} = \frac{2}{4} - \frac{1}{4} = \frac{1}{4} = 2x$

Divide by 2.

$\frac{1}{4} \div 2 = \frac{1}{4} \div \frac{2}{1} = \frac{1}{4} \times \frac{1}{2} = \boxed{\frac{1}{8}} = x$

Check: $3\left(\frac{1}{8}\right) + \frac{1}{2} = \frac{7}{8} = \frac{1}{4} + 5\left(\frac{1}{8}\right)$

⑦ $\frac{x}{2} + 8 = \frac{x}{3} + 6$

Subtract $\frac{x}{3}$ and 8.

$\frac{x}{2} - \frac{x}{3} = 6 - 8$

$\frac{x}{2}\left(\frac{3}{3}\right) - \frac{x}{3}\left(\frac{2}{2}\right) = \frac{3x}{6} - \frac{2x}{6} = \frac{x}{6} = -2$

Multiply by 6.

$x = -2(6) = \boxed{-12}$

Check: $\frac{-12}{2} + 8 = 2 = \frac{-12}{3} + 6$

⑥ $5 - 6x = 7 - 3x$

Add $6x$ and subtract 7.

$5 - 7 = -3x + 6x$

$-2 = 3x$

Divide by 3.

$\boxed{-\frac{2}{3}} = x$

Check: $5 - 6\left(-\frac{2}{3}\right) = 9 = 7 - 3\left(-\frac{2}{3}\right)$

⑧ $\frac{3x}{4} + \frac{1}{2} = \frac{2x}{3} - \frac{5}{6}$

Subtract $\frac{1}{2}$ and $\frac{2x}{3}$.

$\frac{3x}{4} - \frac{2x}{3} = -\frac{5}{6} - \frac{1}{2}$

$\frac{3x}{4}\left(\frac{3}{3}\right) - \frac{2x}{3}\left(\frac{4}{4}\right) = -\frac{5}{6} - \frac{1}{2}\left(\frac{3}{3}\right)$

$\frac{9x}{12} - \frac{8x}{12} = \frac{x}{12} = -\frac{5}{6} - \frac{3}{6} = -\frac{8}{6} = -\frac{4}{3}$

Multiply $\frac{x}{12} = -\frac{4}{3}$ by 12.

$x = -\frac{4}{3}(12) = -\frac{48}{3} = \boxed{-16}$

Check: $\frac{3(-16)}{4} + \frac{1}{2} = -\frac{23}{2} = \frac{2(-16)}{3} - \frac{5}{6}$

Page 114

① $x^{5+3} = x^8$

② $x^{8-7} = x^1 = x$

③ $(x^3)^4 = x^{3(4)} = x^{12}$

④ $(2x^3)^5 = 2^5(x^3)^5 = 32x^{3(5)} = 32x^{15}$

Note: $x = x^1$.

⑤ $4^{3/2} = \left(\sqrt{4}\right)^3 = 2^3 = 8$

⑥ $64^{2/3} = \left(\sqrt[3]{64}\right)^2 = 4^2 = 16$

Notes: $\sqrt[3]{64} = 4$ because $4^3 = (4)(4)(4) = 64$. ($\sqrt[3]{64}$ means the cube root.)

⑦ $\frac{16x^8x^6}{(2x^5)^2} = \frac{16x^{8+6}}{2^2(x^5)^2} = \frac{16x^{14}}{4x^{5(2)}} = \frac{4x^{14}}{x^{10}} = 4x^{14-10} = 4x^4$

⑧ $\frac{x^{12}x^{-8}}{x^{-9}} = \frac{x^{12-8}}{x^{-9}} = \frac{x^4}{x^{-9}} = x^{4-(-9)} = x^{4+9} = x^{13}$

Page 115

    ① $4(3x) - 4(4) = 12x - 16$

    ② $x(x) + x(5) = x^2 + 5x$

    ③ $8x^4(3x^2) + 8x^4(-4x) = 24x^6 - 32x^5$

    ④ $-3(2x) - 3(-6) = -6x + 18$

    ⑤ $5x^3(4x^6) - 5x^3(3x^4) + 5x^3(x^2) = 20x^9 - 15x^7 + 5x^5$

Page 116

    ① $x^2 + 6x + 7x + 42 = x^2 + 13x + 42$

    ② $x^2 - 5x + 5x - 25 = x^2 - 25$

    ③ $x^2 + 5x + 5x + 25 = x^2 + 10x + 25$

    ④ $8x^2 - 24x + 6x - 18 = 8x^2 - 18x - 18$

    ⑤ $x^3 + 5x^2 - 4x^2 - 20x = x^3 + x^2 - 20x$

    ⑥ $3x(4x) + 3x(-y) + 2y(4x) + 2y(-y) = 12x^2 - 3xy + 8xy - 2y^2$

    $= 12x^2 + 5xy - 2y^2$

Page 117

    ① $12x^2 - 18 = 6(2x^2 - 3)$

    ② $8x^3 + 12x^2 = 4x^2(2x + 3)$

    ③ $27x^4 - 36x = 9x(3x^3 - 4)$

    ④ $-x^{12} - x^{10} = -x^{10}(x^2 + 1)$

    ⑤ $9x^7 - 6x^5 + 3x^3 = 3x^3(3x^4 - 2x^2 + 1)$

    Note: You can check these answers by distributing.

Page 118

    ① $8x\sqrt{x} - 12x$

    ② $\sqrt{3x}\sqrt{x} - \sqrt{3x}\sqrt{3} = \sqrt{3x^2} - \sqrt{9x} = x\sqrt{3} - 3\sqrt{x}$

    ③ $2\sqrt{x} - x + 6 - 3\sqrt{x} = -x + (2 - 3)\sqrt{x} + 6 = -x - \sqrt{x} + 6$

    ④ $4\sqrt{x}(3\sqrt{x} - 2)$

    ⑤ $x\sqrt{x}(x + 1)$

    Note to Problems 4-5: You can check these answers by distributing.

Page 119

① $\dfrac{1}{x} - \dfrac{2}{3} = \dfrac{1}{12}$

$\dfrac{1}{x} = \dfrac{1}{12} + \dfrac{2}{3}$

$\dfrac{1}{x} = \dfrac{1}{12} + \dfrac{2}{3}\left(\dfrac{4}{4}\right) = \dfrac{1}{12} + \dfrac{8}{12} = \dfrac{9}{12} = \dfrac{3}{4}$

Take the reciprocal of $\dfrac{1}{x} = \dfrac{3}{4}$.

$x = \boxed{\dfrac{4}{3}}$

Check: $\dfrac{1}{4/3} - \dfrac{2}{3} = \dfrac{3}{4} - \dfrac{2}{3} = \dfrac{1}{12}$

Alternative: cross multiply.

② $\dfrac{4}{3x} + \dfrac{1}{36} = \dfrac{3}{2x}$

$\dfrac{1}{36} = \dfrac{3}{2x} - \dfrac{4}{3x}$

$\dfrac{1}{36} = \dfrac{3}{2x}\left(\dfrac{3}{3}\right) - \dfrac{4}{3x}\left(\dfrac{2}{2}\right) = \dfrac{9}{6x} - \dfrac{8}{6x} = \dfrac{1}{6x}$

$36 = 6x$

$\dfrac{36}{6} = \boxed{6} = x$

Check: $\dfrac{4}{3(6)} + \dfrac{1}{36} = \dfrac{4}{18} + \dfrac{1}{36} = \dfrac{9}{36} = \dfrac{1}{4}$

$\dfrac{3}{2x} = \dfrac{3}{2(6)} = \dfrac{3}{12} = \dfrac{1}{4}$

Page 120

① $72 = 4x$ $\rightarrow$ $\dfrac{72}{4} = \boxed{18} = x$ Check: $\dfrac{18}{8} = \dfrac{9}{4}$

② $8x = 168$ $\rightarrow$ $x = \dfrac{168}{8} = \boxed{21}$ Check: $\dfrac{24}{21} = \dfrac{8}{7}$

③ $360 = 15x$ $\rightarrow$ $\dfrac{360}{15} = \boxed{24} = x$ Check: $\dfrac{18}{24} = \dfrac{3}{4} = \dfrac{3}{4}\left(\dfrac{5}{5}\right) = \dfrac{15}{20}$

④ $6x = 72$ $\rightarrow$ $x = \dfrac{72}{6} = \boxed{12}$ Check: $\dfrac{12}{9} = \dfrac{4}{3} = \dfrac{4}{3}\left(\dfrac{2}{2}\right) = \dfrac{8}{6}$

Page 121

① $3x = 3x$ is satisfied by all real numbers

② $4x - 4x = 8 - 6$ becomes $0 = 2$ so there is no solution

③ $x^2 - 7x =$ factors as $x(x - 7) = 0$ with two answers: $x = 0$ or $x = 7$

Check: $0^2 = 7(0)$ and $7^2 = 7(7)$

④ $x + 5 = 5 + x$ is satisfied by all real numbers

⑤ $2x + 1 = 2x + 5$ becomes $1 = 5$ so there is no solution

⑥ Cross multiply: $12x = 2x^2$ becomes $6x = x^2$ or $0 = x^2 - 6x$, which factors as $0 = x(x - 6)$ with two answers: $x = 0$ or $x = 6$

Check: $\dfrac{3(0)}{2} = \dfrac{0^2}{4}$ and $\dfrac{3(6)}{2} = \dfrac{18}{2} = 9 = \dfrac{6^2}{4} = \dfrac{36}{4}$

Page 122

There aren't any problems on page 122.

Page 123

① $4x > 12$

$x > \frac{12}{4}$

$x > 3$

③ $6 < 2x$

$\frac{6}{2} < x$

$3 < x$

⑤ $0 > 9 + 9x$

$-9 > 9x$

$-1 > x$

Alternatively, $x < -1$

⑦ $x < -20$

Alternatively, $-20 > x$

② $x > -9$ (reverse when dividing by -1)

Alternatively: $0 < 9 + x$ such that $-9 < x$

Note: $x > -9$ is equivalent to $-9 < x$

④ $-3x > 6$

$x < -2$ (reverse when dividing by -3)

Alternatively: $3 - 9 > 3x$ such that $-6 > 3x$

and $-2 > x$ (equivalent to $x < -2$)

⑥ $4x - 3x < 2 - 8$

$x < -6$

Alternatively, $-6 > x$

⑧ $70 > -5x$

$-14 < x$ (reverse when dividing by -5)

Alternatively: $70 + 5x > 0$ such that $5x > -70$

and $x > -14$ (equivalent to $-14 < x$)

Page 124

① (B) 3

② (A) $x^2 + 4x - 3$ since there is no equal sign

③ (E) $x - 7$

④ (A) $\frac{y}{5}$

⑤ (E) $x - y$

⑥ (E) $a = 8$ since $a = 2(3)^2 - 5(3) + 5 = 2(9) - 15 + 5 = 18 - 10 = 8$

⑦ (C) $C = 8$ since $C = \frac{Q}{V} = \frac{24}{3} = 8$

⑧ (D) $5x - 10$ since $9x - 3 - 5x + 1 + x - 8 = (9 - 5 + 1)x - 3 + 1 - 8 =$

$5x - 11 + 1 = 5x - 10$

Note: $x = 1x$

Page 125

⑨ (B) $x = -\frac{7}{4}$ since $2 - 4x = 9 \rightarrow -4x = 9 - 2 \rightarrow -4x = 7 \rightarrow x = -\frac{7}{4}$

Check: $2 - 4\left(-\frac{7}{4}\right) = 2 + 7 = 9$

⑩ (C) $y = 8$ since $5y - 12 = 36 - y \rightarrow 5y + y = 36 + 12 \rightarrow 6y = 48 \rightarrow$

$y = \frac{48}{6} = 8$

Check: $5y - 12 = 5(8) - 12 = 40 - 12 = 28$ and $36 - y = 36 - 8 = 28$

⑪ (A) $2x^2 - 7x - 15$ since $(x - 5)(2x + 3) = 2x^2 + 3x - 10x - 15$

$= 2x^2 - 7x - 15$ (using the f.o.i.l. method)

⑫ (E) $8x^5 - 20x$ since $4x(2x^4 - 5) = 4x(2x^4) + 4x(-5) = 8x^5 - 20x$

⑬ (D) $4x^3(3x^2 - 4)$ since $12x^5 - 16x^3 = 4x^3(3x^2) - 4x^3(4) = 4x^3(3x^2 - 4)$

⑭ (C) $4x$ since $\sqrt{2x}\sqrt{8x} = \sqrt{(2x)(8x)} = \sqrt{16x^2} = \sqrt{16}\sqrt{x^2} = 4x$

Note: $\sqrt{ab} = \sqrt{a}\sqrt{b}$ (technically there is also a negative root: $\pm 4x$)

⑮ (A) $x = \frac{1}{3}$ since $\frac{3}{x} = 9 \rightarrow 3 = 9x \rightarrow \frac{3}{9} = x \rightarrow \frac{1}{3} = x$ (cross multiplying)

Check: $\frac{3}{1/3} = 3 \div \frac{1}{3} = \frac{3}{1} \div \frac{1}{3} = \frac{3}{1} \times \frac{3}{1} = \frac{9}{1} = 9$

⑯ (C) $x = 18$ since $\frac{48}{x} = \frac{8}{3}$ (cross multiply) $\rightarrow 48(3) = 8x \rightarrow 144 = 8x$

$\rightarrow \frac{144}{8} = x \rightarrow 18 = x$

Check: $\frac{48}{18} = \frac{48 \div 6}{18 \div 6} = \frac{8}{3}$

Page 126

⑰ (B) $L < 5$ is the best answer.

⑱ (D) $x < 3$ since $6x - 9 < 9 \rightarrow 6x < 9 + 9 \rightarrow 6x < 18 \rightarrow x < \frac{18}{6} \rightarrow x < 3$

⑲ (A) $x < -2$ since $2 - 3x > 8 \rightarrow -3x > 6 \rightarrow x < -2$

Note: Change $>$ into $<$ when dividing by negative 3.

Alternatively, $2 - 3x > 8 \rightarrow 2 > 8 + 3x \rightarrow -6 > 3x \rightarrow -2 > x$, which is equivalent to $x < -2$.

⑳ (E) $x > -6$ is equivalent to $6 > -x$ and $-6 < x$.

Note: Change $>$ into $<$ when dividing by negative 1.

Alternatively, $6 > -x \rightarrow 6 + x > 0 \rightarrow x > -6$.

Chapter 6: Relationships

Page 127

There aren't any problems on page 127.

Page 128

① Add 273 to Celsius to get Kelvin. Example: $70 + 273 = 343$

Celsius	60	70	80	90	100
Kelvin	333	343	353	363	373

② Subtract 2.4 from the price to get the profit. Example: $4.2 - 1.8 = 2.4$

price	4.2	4.5	4.8	5.1	5.4
profit	1.8	2.1	2.4	2.7	3

③ Add $\frac{1}{4}$ to the width to get the gap. $\frac{1}{16} + \frac{1}{4} = \frac{1}{16} + \frac{4}{16} = \frac{5}{16}, \frac{1}{8} + \frac{1}{4} = \frac{1}{8} + \frac{2}{8} = \frac{3}{8}$,

$\frac{3}{16} + \frac{1}{4} = \frac{3}{16} + \frac{4}{16} = \frac{7}{16}, \frac{1}{4} + \frac{1}{4} = \frac{2}{4} = \frac{1}{2}$, and $\frac{5}{16} + \frac{1}{4} = \frac{5}{16} + \frac{4}{16} = \frac{9}{16}$

width	1/16	1/8	3/16	1/4	5/16
gap	5/16	3/8	7/16	1/2	9/16

Alternatively, convert the fractions to decimals and add 0.25.

Example: $0.0625 + 0.25 = 0.3125$

width	0.0625	0.125	0.1875	0.25	0.3125
gap	0.3125	0.375	0.4375	0.5	0.5625

Page 129

① Multiply the distance by 12 to get the time. Example: $(6)(12) = 72$

distance	6	12	18	24	30
time	72	144	216	288	360

② Multiply Amps by 2.4 to get Volts. Example: $(3)(2.4) = 7.2$

Amps	3	4.5	6	7.5	9
Volts	7.2	10.8	14.4	18	21.6

③ Multiply force by 4 to get work. Example: $\frac{1}{72}\frac{4}{1} = \frac{4}{72} = \frac{1}{18}$

force	1/72	1/36	1/24	1/18	5/72
work	1/18	1/9	1/6	2/9	5/18

Page 130

① Multiplicative. Each value of E is 2 times f. Example: $(5)(2) = 10$

② Additive. Each value of z is 4 more than w. Example: $20 + 4 = 24$

③ Multiplicative. Each value of E is $\frac{1}{5}$ times S. Example: $\frac{10.25}{5} = 10.25 \div 5 = 2.05$

Page 131

① $L = 4p$ Example: $(45)(4) = 180$

② $c = b - 2$ Example: $8 - 6 = 2$

③ $R = 0.5D = \frac{D}{2} = D \div 2$ Example: $\frac{0.17}{2} = 0.085$

Page 132

① $U = 12C = 12(8) = \boxed{96}$

② $E = H + 1.2 = 8.8 + 1.2 = \boxed{10}$

③ $T = 60F = 60(400) = \boxed{24{,}000}$

④ $j = k - \frac{1}{4} = \frac{7}{12} - \frac{1}{4} = \frac{7}{12} - \frac{1(3)}{4(3)} = \frac{7}{12} - \frac{3}{12} = \frac{4}{12} = \frac{4 \div 4}{12 \div 4} = \boxed{\frac{1}{3}}$

⑤ $I = 0.15P = 0.15(\$1200) = \frac{15}{100}(\$1200) = 15\frac{1200}{100} = 15(12) = \boxed{\$180}$

Page 133

① Multiply by 4 and add 2. Examples: $8(4) + 2 = 34$, $12(4) + 2 = 50$

red	8	12	16	20	24
green	34	50	66	82	98

② Multiply by 0.02 and add 1. Example: $150(0.02) + 1 = 6 + 1 = 7$

Alternative: Divide by 25 and add 1. Example: $150 \div 25 + 1 = 6 + 1 = 7$

students	150	175	200	225	250
teachers	7	8	9	10	11

③ Multiply by 5 and subtract 0.1. Example: $0.4(5) - 0.1 = 2 - 0.1 = 1.9$

weight	0.4	0.7	1	1.3	1.6
price	1.9	3.4	4.9	6.4	7.9

Page 134

There aren't any problems on page 134.

Page 135

① The coefficient is: $\frac{v_2-v_1}{t_2-t_1} = \frac{2.1-1.5}{0.9-0.6} = \frac{0.6}{0.3} = 2$

$\boxed{v = 2t + 0.3}$ Examples: $2(0.6) + 0.3 = 1.5$ and $2(0.9) + 0.3 = 2.1$

Alternate answer: $t = 0.5v - 0.15$

② The coefficient is: $\frac{R_2-R_1}{Q_2-Q_1} = \frac{103-75}{15-11} = \frac{28}{4} = 7$

$\boxed{R = 7Q - 2}$ Examples: $7(11) - 2 = 77$ and $7(15) - 2 = 103$

Alternate answer: $Q = \frac{R+2}{7}$

③ The coefficient is: $\frac{U_2-U_1}{T_2-T_1} = \frac{455-215}{80-40} = \frac{240}{40} = 6$

$\boxed{U = 6T - 25}$ Examples: $6(40) - 25 = 215$ and $6(80) - 25 = 455$

Alternate answer: $T = \frac{U+25}{6}$

Page 136

① $P = 0.4(7) - 1.2 = 2.8 - 1.2 = \boxed{1.6}$

② We are given $u = 14$ and are predicting t. We must isolate t first.

Add 1 to get $u + 1 = 3t$ and divide by 3 to get $\frac{u+1}{3} = t$

$t = \frac{u+1}{3} = \frac{14+1}{3} = \frac{15}{3} = \boxed{5}$ Check: $u = 3t - 1 = 3(5) - 1 = 15 - 1 = 14$

③ $F = 1.8C + 32 = 1.8(100) + 32 = 180 + 32 = 212$

Page 137

① Q and V are directly proportional

② P and R are inversely proportional

③ M and p are inversely proportional

④ v and f are directly proportional

Note: $v = \frac{f}{2} = 0.5f$ since $\frac{1}{2} = 0.5$. Example: $\frac{8}{2} > \frac{4}{2}$ since $4 > 2$

⑤ W and F are directly proportional

⑥ x and y are directly proportional

Note: Cross multiply (Sec. 5.15) to get $y = x$.

⑦ n and v are inversely proportional

Note: Divide both sides by v to get $n = \frac{3}{v}$.

⑧ V and R are directly proportional

Note: Multiply both sides by R to get $V = 0.4R$.

Example: $0.4(9) > 0.4(3)$ since $3.6 > 1.2$

Page 138

① $4(128) = 512$ so top row times bottom row equals 512

Divide 512 by any number to get its counterpart. Example: $512 \div 16 = 32$

time	4	8	16	32	64
rate	128	64	32	16	8

② $0.08(125) = 10$ so top row times bottom row equals 10

Divide 10 by any number to get its counterpart. Example: $10 \div 25 = 0.4$

Amps	2	0.4	0.08	0.016	0.0032
Ohms	5	25	125	625	3125

③ $6(16) = 96$ so top row times bottom row equals 96

Divide 96 by any number to get its counterpart. Example: $96 \div 2 = 48$

base	2	3	4	5	6
height	48	32	24	19.2	16

Page 139

① The constant is $(5)(800) = 4000$ such that $\boxed{V = \frac{4000}{d}}$

Example: $\frac{4000}{10} = 400$ Alternate answer: $dV = 4000$

② The constant is $(0.002)(30{,}000) = (2)(30) = 60$ such that $\boxed{P = \frac{60}{A}}$

Example: $\frac{60}{0.02} = 3000$ Alternate answer: $PA = 60$

③ The constant is $(560)\left(\frac{1}{7}\right) = 80$ such that $\boxed{W = \frac{80}{L}}$

Example: $\frac{80}{560} = \frac{1}{7}$ Alternate answer: $LW = 80$

Page 140

① $I = \dfrac{2}{R} = \dfrac{2}{5} = \dfrac{2 \times 2}{5 \times 2} = \dfrac{4}{10} = \boxed{0.4}$ Check $IR = (0.4)(5) = 2$

② Multiply by t to get $rt = 1200$. Divide by r to get $t = \dfrac{1200}{r}$.

$t = \dfrac{1200}{r} = \dfrac{1200}{60} = \boxed{20}$ Check $rt = (60)(20) = 1200$

③ Divide by b to get $h = \dfrac{3.6}{b}$.

$h = \dfrac{3.6}{0.8} = \dfrac{3.6 \times 10}{0.8 \times 10} = \dfrac{36}{8} = \dfrac{36 \div 4}{8 \div 4} = \dfrac{9}{2} = \boxed{4.5}$ Check $bh = (0.8)(4.5) = 3.6$

Page 141

① distributive property

② commutative property of addition: $-1 + x = x + (-1) = x - 1$

③ Step 1 is the associative property of addition: $1 + x - 1 = 1 - 1 + x$

Step 2 is the inverse property of addition: $1 + (-1) = 1 - 1 = 0$

Step 3 is the identity property of addition: $x + 0 = x$

④ Step 1 is the commutative property of multiplication: $(x + 2)3 = 3(x + 2)$

Step 2 is the distributive property: $(x + 2)3 = 3x + 6$

⑤ associative property of multiplication: $4(2 - x)2 = 4(2)(2 - x) = 8(2 - x)$

Note: Associative involves three factors, whereas commutative involves two.

⑥ associative property of multiplication: $\dfrac{1}{2} \dfrac{1}{3} \dfrac{1}{x} = \dfrac{1}{2} \dfrac{1}{3} \dfrac{1}{x} = \dfrac{1}{6x}$

Page 142

① yes: $8x - 2 - 5x + 7 = (8x - 5x) + (7 - 2) = 3x + 5$

② no: $\dfrac{1}{4}(4x - 1) = \dfrac{1}{4}(4x) - \dfrac{1}{4}(1) = x - \dfrac{1}{4} \neq x - 1$

③ yes: $2x - 3 + 4x - 5 = (2x + 4x) + (-3 - 5) = 6x - 8 = 2(3x) - 2(4)$

$= 2(3x - 4)$

④ yes: $5x^2 - 4x + 3x^2 = (5 + 3)x^2 - 4x = 8x^2 - 4x = 4x(2x) - 4x(1)$

$= 4x(2x - 1)$

Page 143

There aren't any problems on page 143.

Page 144

① Let $x =$ the amount of money that Sarah had in the beginning.

Model: $x - \$14.28 = \$19.85$

Solution: $x = \$19.85 + \$14.28 = \boxed{\$34.13}$ (Sarah originally had \$34.13)

② Let $x =$ Cindy's age.

Model: $15 = 5x$

Solution: $\frac{15}{5} = \boxed{3} = x$ (Cindy is 3 years old)

③ Let $x =$ the smaller number, such that $2x =$ the larger number

Model: $x + 2x = 36$

Solution: $3x = 36 \rightarrow x = \frac{36}{3} = 12$ (the numbers are $x = \boxed{12}$ and $2x = \boxed{24}$)

Page 145

① (D) 73 (add 7: $38 + 7 = 45, 66 + 7 = 73$)

② (C) 128 (multiply by 2: $4 \times 2 = 8, 64 \times 2 = 128$)

③ (B) multiplicative: $u = 3t$ ($4 \times 3 = 12, 6 \times 3 = 18$)

④ (A) additive: $G = F + 3$ ($3 + 3 = 6, 9 + 3 = 12$)

⑤ (B) $q = 2p$ ($12 \times 2 = 24, 15 \times 2 = 30$)

⑥ (D) $c = b + 5$ ($15 + 5 = 20, 30 + 5 = 35$)

Page 146

⑦ (A) $w = 1$ since $w = \frac{z}{2} - 3 = \frac{8}{2} - 3 = 4 - 3 = 1$

⑧ (A) $h = 7$ since $b = 3h + 1 = 3(7) + 1 = 21 + 1 = 22$

Note: $b - 1 = 3h \rightarrow \frac{b-1}{3} = h \rightarrow \frac{22-1}{3} = h \rightarrow \frac{21}{3} = h \rightarrow 7 = h$

⑨ (D) $y = 4x + 2$ ($4 \times 3 + 2 = 12 + 2 = 14, 4 \times 5 + 2 = 20 + 2 = 22$)

⑩ (B) inverse since $y = \frac{5}{x}$

⑪ (A) direct

⑫ (C) distributive

⑬ (C) $2(x - 3)$ since $9x + 1 - 7x - 7 = (9x - 7x) + (1 - 7) = 2x - 6 = 2(x - 3)$

⑭ (A) $x - 5 = 9$ (such that $x = 9 + 5 = 14$)

⑮ (C) $5x = 8$ (such that $x = \frac{8}{5} = 1.6$)

Chapter 7: Data Analysis

Page 147

There aren't any problems on page 147.

Page 148

① $\frac{12+13+15+16}{4} = \frac{56}{4} = \boxed{14}$ Check: The answer lies between 12 and 16.

② $\frac{4+6+7+8+9}{5} = \frac{34}{5} = \frac{34\times2}{5\times2} = \frac{68}{10} = \boxed{6.8}$ Check: The answer lies between 4 and 9.

Page 149

① Put the values in order: 9, 11, 14, 15, 16. The middle value (median) is $\boxed{14}$.

The mean is $\frac{9+11+14+15+16}{5} = \frac{65}{5} = \boxed{13}$.

② Put the values in order: 1, 2, 4, 6, 8, 9. The middle values are 4 and 6. The mean of the middle values is the median: $\frac{4+6}{2} = \frac{10}{2} = \boxed{5}$.

The mean is $\frac{1+2+4+6+8+9}{6} = \frac{30}{6} = \boxed{5}$. (They happen to be equal this time.)

Page 150

① The range is $14 - 4 = \boxed{10}$.

Put the values in order: 4, 6, 8, 9, 13, 14. The middle values are 8 and 9. The mean of the middle values is the median: $\frac{8+9}{2} = \frac{17}{2} = \boxed{8.5}$.

The mean is $\frac{4+6+8+9+13+14}{6} = \frac{54}{6} = \boxed{9}$.

② The range is $1.3 - 0.5 = \boxed{0.8}$.

Put the values in order: 0.5, 0.9, 1.1, 1.3. The middle values are 0.9 and 1.1.

The mean of the middle values is the median: $\frac{0.9+1.1}{2} = \frac{2}{2} = \boxed{1}$.

The mean is $\frac{0.5+0.9+1.1+1.3}{4} = \frac{3.8}{4} = \frac{3.8\times25}{4\times25} = \frac{95}{100} = \boxed{0.95}$.

Page 151

There aren't any problems on page 151.

Page 152

① The mean is $\frac{24+27+21}{3} = \frac{72}{3} = \boxed{24}$. The standard deviation is:

$$\text{std. dev.} = \sqrt{\frac{(24-24)^2 + (27-24)^2 + (21-24)^2}{3-1}}$$

$$\text{std. dev.} = \sqrt{\frac{(0)^2 + (3)^2 + (-3)^2}{2}} = \sqrt{\frac{0+9+9}{2}} = \sqrt{\frac{18}{2}} = \sqrt{9} = \boxed{3}$$

Note that $(-3)^2 = (-3) \times (-3) = 9$ (Sec. 1.3).

The range is $27 - 21 = \boxed{6}$.

Put the values in order: 21, 24, 27. The middle value (median) is $\boxed{24}$.

② The mean is $\frac{4+10+4+5+2}{5} = \frac{25}{5} = \boxed{5}$. The standard deviation is:

$$\text{std. dev.} = \sqrt{\frac{(4-5)^2 + (10-5)^2 + (4-5)^2 + (5-5)^2 + (2-5)^2}{5-1}}$$

$$\text{std. dev.} = \sqrt{\frac{(-1)^2 + (5)^2 + (-1)^2 + (0)^2 + (-3)^2}{4}} = \sqrt{\frac{1+25+1+0+9}{4}}$$

$$\text{std. dev.} = \sqrt{\frac{36}{4}} = \sqrt{9} = \boxed{3}$$

Note that $(-1)^2 = (-1) \times (-1) = 1$ and $(-3)^2 = (-3) \times (-3) = 9$ (Sec. 1.3).

The range is $10 - 2 = \boxed{8}$.

Put the values in order: 2, 4, 4, 5, 10. The middle value is $\boxed{4}$.

Page 153

There aren't any problems on page 153.

Page 154

① Put the values in order: 47, 48, 48, 50, 51, 54, 54, 59, 59, 61, 62, 63.

The middle values are 54 and 54. The median is $\frac{54+54}{2} = \frac{108}{2} = \boxed{54}$.

The range is $63 - 47 = \boxed{16}$.

The lower half includes 47, 48, 48, 50, 51, 54. The LQ is $\frac{48+50}{2} = \frac{98}{2} = 49$.

The upper half includes 54, 59, 59, 61, 62, 63. The UQ is $\frac{59+61}{2} = \frac{120}{2} = 60$.

The interquartile range is IQR $=$ UQ $-$ LQ $= 60 - 49 = \boxed{11}$.

② Put the values in order: 3.6, 3.8, 3.9, 3.9, 4, 4.1, 4.1, 4.2, 4.2, 4.3.

The middle values are 4 and 4.1. The median is $\frac{4+4.1}{2} = \frac{8.1}{2} = \boxed{4.05}$.

The range is $4.3 - 3.6 = \boxed{0.7}$.

The lower half includes 3.6, 3.8, 3.9, 3.9, 4. The LQ is 3.9.

The upper half includes 4.1, 4.1, 4.2, 4.2, 4.3. The UQ is 4.2.

The interquartile range is IQR $=$ UQ $-$ LQ $= 4.2 - 3.9 = \boxed{0.3}$.

Page 155

There aren't any problems on page 155.

Page 156

① Order the data. Lower half: 12, 13, 14, 14. Upper half: 15, 16, 16, 17.

LV $= 12$, GV $= 17$, MD $= \frac{14+15}{2} = \frac{29}{2} = 14.5$, LQ $= \frac{13+14}{2} = \frac{27}{2} = 13.5$, UQ $= 16$

The interquartile range is IQR $=$ UQ $-$ LQ $= 16 - 13.5 = \boxed{2.5}$.

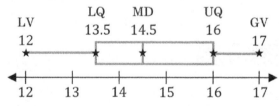

② Order the data. Lower half: 40, 42, 42, 44, 44, 44.

Upper half: 44, 47, 47, 49, 49, 50.

LV $= 40$, GV $= 50$, MD $= 44$, LQ $= \frac{42+44}{2} = \frac{86}{2} = 43$, UQ $= \frac{47+49}{2} = \frac{96}{2} = 48$

The interquartile range is IQR $=$ UQ $-$ LQ $= 48 - 43 = \boxed{5}$.

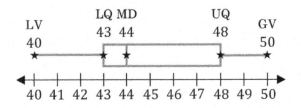

Page 157

① The frequency of 5's is ☐1.

The frequency of 6's is ☐2.

The frequency of 7's is ☐5.

The frequency of 8's is ☐3.

The frequency of 9's is ☐2.

The frequency of 10's is ☐1.

Check: There are $1 + 2 + 5 + 3 + 2 + 1 = 14$ data values.

Page 158

① Check: There are $1 + 2 + 4 + 3 + 2 + 1 + 0 + 1 = 14$ data values.

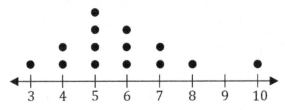

Page 159

There aren't any problems on page 159.

Page 160

① Check: There are $1 + 2 + 4 + 6 + 2 = 15$ data values.

Int.	Freq.
1-2	1
3-4	2
5-6	4
7-8	6
9-10	2

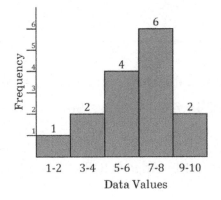

② Check: There are $1 + 3 + 5 + 3 + 1 = 13$ data values.

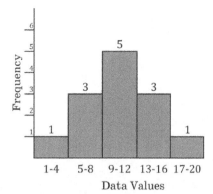

Int.	Freq.
1-4	1
5-8	3
9-12	5
13-16	3
17-20	1

Page 161

① First organize the data into groups by tens digits.

Seventies: 75, 77, 78. Eighties: 80, 83, 84, 87, 89. Nineties: 90, 91, 95, 95, 97.

Stems	Leaves
7	5 7 8
8	0 3 4 7 9
9	0 1 5 5 7

Page 162

There aren't any problems on page 162.

Page 163

① There are $3 + 2 + 4 + 4 + 3 = 16$ data values.

Count the dots in each category to determine the frequencies.

Divide each frequency by 16 to determine the relative frequency.

Convert the fraction to a decimal (Sec. 3.9) and convert the decimal to a percent (Sec. 3.7). Note that $\frac{3}{16} = \frac{3 \times 625}{16 \times 625} = \frac{1875}{10,000} = 18.75\%$,

$\frac{2}{16} = \frac{2 \div 2}{16 \div 2} = \frac{1}{8} = \frac{125}{8 \times 125} = \frac{125}{1000} = 12.5\%$, and $\frac{4}{16} = \frac{4 \div 4}{16 \div 4} = \frac{1}{4} = \frac{1 \times 25}{4 \times 25} = \frac{25}{100} = 25\%$.

CO_2	H_2O	NH_3	CH_4	N_2O
3	2	4	4	3
$\frac{3}{16} = 18.75\%$	$\frac{2}{16} = 12.5\%$	$\frac{4}{16} = 25\%$	$\frac{4}{16} = 25\%$	$\frac{3}{16} = 18.75\%$

There are two modes: $\boxed{NH_3}$ and $\boxed{CH_4}$ are tied with the highest frequency (4).

Page 164

There aren't any problems on page 164.

Page 165

① Label 5% increments on the vertical axis. The mode is pants (40%).

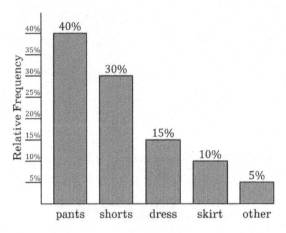

Page 166

① Baseball (50%) should take half the pie, soccer (25%) should take one-fourth of the pie, and the remainder should be split evenly between tennis (12.5%) and golf (12.5%).

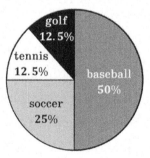

Page 167

There aren't any problems on page 167.

Page 168

① The range is $7 - 1 = \boxed{6}$.

The data values are: 1, 2, 2, 3, 3, 3, 3, 4, 4, 4, 4, 4, 4, 5, 5, 5, 5, 6, 6, 7.

The median is $\boxed{4}$.

The mean is $\dfrac{1+2+2+3+3+3+3+4+4+4+4+4+4+5+5+5+5+6+6+7}{1+2+4+6+4+2+1}$ (see the next page)

$$= \frac{1+2(2)+3(4)+4(6)+5(4)+6(2)+7(1)}{20} = \frac{1+4+12+24+20+12+7}{20} = \frac{80}{20} = \boxed{4}.$$

② The data is symmetric, clustered from 1 to 7, and has a peak at 4 with no apparent outliers.

Page 169

① (C) 1.6 since $\frac{2.1+1.3+0.8+1.5+2.3+1.6}{6} = \frac{9.6}{6} = 1.6$

② (D) 1.55 since ordering the data gives 0.8, 1.3, 1.5, 1.6, 2.1, 2.3 and since $\frac{1.5+1.6}{2} = \frac{3.1}{2} = 1.55$

③ (D) 1.5 since range = GV − LV = 2.3 − 0.8 = 1.5

④ (B) 0.8 since the lower half if 0.8, 1.3, 1.5 and the upper half is 1.6, 2.1, 2.3. The lower quartile (LQ) is 1.3 and the upper quartile (UQ) is 2.1.

IQR = UQ − LQ = 2.1 − 1.3 = 0.8

⑤ (B) 14

⑥ (C) 8 since range = GV − LV = 19 − 11 = 8

⑦ (D) 4 since IQR = UQ − LQ = 17 − 13 = 4

Page 170

⑧ (C) 2.8 since $\frac{1+2+2+2+2+3+3+4+4+5}{1+4+2+2+1} = \frac{1+2(4)+3(2)+4(2)+5}{10} = \frac{1+8+6+8+5}{10} = \frac{28}{10} = 2.8$

⑨ (B) 2.5 since ordering the data gives 1, 2, 2, 2, 2, 3, 3, 4, 4, 5 and since $\frac{2+3}{2} = \frac{5}{2} = 2.5$

⑩ (D) 4 since range = GV − LV = 5 − 1 = 4

⑪ (D) 4 since the data value 2 occurs 4 times

⑫ (D) pennies since pennies has the greatest frequency.

Note: The mode is the result that occurs most often. (It would be incorrect to give 24 as the answer because 24 isn't a result that occurs. Pennies is a result that occurs 24 times.)

⑬ (B) 25% since the total number is 24 + 12 + 9 + 3 = 48 and since $\frac{12}{48} = \frac{12 \div 12}{48 \div 12} = \frac{1}{4} = \frac{1 \times 25}{4 \times 25} = \frac{25}{100} = 25\%$

⑭ (B) $2.49 since 24 pennies makes $24 \times \$0.01 = \$0.24$, 12 nickels makes $12 \times \$0.05 = \$0.60$, 9 dimes makes $9 \times \$0.10 = \$0.90$, 3 quarters makes $3 \times \$0.25 = \$0.75$, and the total is $\$0.24 + \$0.60 + \$0.90 + \$0.75 = \$2.49$.

Page 171

⑮ (B) 82.5 since $\dfrac{58+72+75+75+78+79+80+83+84+84+87+89+91+91+96+98}{16} = \dfrac{1320}{16} = 82.5$

⑯ (D) 83.5 since the middle values are 83 and 84, and since $\dfrac{83+84}{2} = \dfrac{167}{2} = 83.5$

⑰ (E) 40 since the least value is 58 and the greatest value is 98, and since range $= GV - LV = 98 - 58 = 40$

⑱ (D) 4 since the top of the 7-9 bar comes to 4 on the vertical axis

Page 172

⑲ (D) March since March has the greatest relative frequency

Note: The mode is the result that occurs most often. (It would be incorrect to give 35% as the answer because 35% isn't a result that occurs. March is a result that occurs 35% of the time.)

⑳ (B) 20 since 25% of the people answered April and since $25\% \times 80$

$= \dfrac{1}{4} \times 80 = \dfrac{80}{4} = 20$ (Note that 20 out of 80 equates to 1/4, which is 25%)

Chapter 8: Coordinate Graphs

Page 173

There aren't any problems on page 173.

Page 174

There aren't any problems on page 174.

Page 175

① $(8, 3)$ ② $(0, 7)$ ③ $(6, 9)$ ④ $(4, 0)$

Page 176

There aren't any problems on page 176.

Page 177

① $(-1, 3)$ left and up   ② $(-2, -5)$ left and down

③ $(3, -4)$ right and down   ④ $(-2, 0)$ left only

Page 178

There aren't any problems on page 178.

Page 179

① $x_1 = 4, y_1 = 2, x_2 = 6, y_2 = 18$: slope $= \frac{y_2 - y_1}{x_2 - x_1} = \frac{18 - 2}{6 - 4} = \frac{16}{2} = \boxed{8}$

② $x_1 = -3, y_1 = 4, x_2 = 5, y_2 = 8$: slope $= \frac{y_2 - y_1}{x_2 - x_1} = \frac{8 - 4}{5 - (-3)} = \frac{4}{5 + 3} = \frac{4}{8} = \boxed{\frac{1}{2}} = \boxed{0.5}$

③ $x_1 = -4, y_1 = 8, x_2 = 4, y_2 = -8$: slope $= \frac{y_2 - y_1}{x_2 - x_1} = \frac{-8 - 8}{4 - (-4)} = \frac{-16}{4 + 4} = \frac{-16}{8} = \boxed{-2}$

④ $x_1 = -7, y_1 = -5, x_2 = 9, y_2 = -5$: slope $= \frac{y_2 - y_1}{x_2 - x_1} = \frac{-5 - (-5)}{9 - (-7)} = \frac{-5 + 5}{9 + 7} = \frac{0}{16} = \boxed{0}$

Note: Since the y-value didn't change (it is -5 initially and finally), it should make sense that the slope is zero (it is a horizontal line).

Page 180

There aren't any problems on page 180.

Page 181

① The endpoints are $(2, 0)$ and $(4, 10)$.

Note that $x_1 = 2, y_1 = 0, x_2 = 4$, and $y_2 = 10$.

slope $= \frac{y_2 - y_1}{x_2 - x_1} = \frac{10 - 0}{4 - 2} = \frac{10}{2} = \boxed{5}$ (rises 5 up for every 1 right)

② The endpoints are $(0, 7)$ and $(10, 2)$.

Note that $x_1 = 0, y_1 = 7, x_2 = 10$, and $y_2 = 2$.

slope $= \frac{y_2 - y_1}{x_2 - x_1} = \frac{2-7}{10-0} = \frac{-5}{10} = \boxed{-\frac{1}{2}} = \boxed{-0.5}$ (drops 1 for every 2 right)

Page 182

There aren't any problems on page 182.

Page 183

① The y-intercept is 8 (where the line crosses the vertical axis).

② The y-intercept is 0.

③ The y-intercept is -6.

④ The y-intercept is 30.

Page 184

There aren't any problems on page 184.

Page 185

① Compare $y = x - 1$ with $y = mx + b$ to see that the slope is $m = \boxed{1}$ and the y-intercept is $b = \boxed{-1}$. Note that $x = 1x$.

② First isolate y. Subtract $4x$ from both sides to get $\frac{y}{3} = -4x + 2$. Multiply both sides by 3 to get $y = -12x + 6$. Compare $y = -12x + 6$ with $y = mx + b$ to see that the slope is $m = \boxed{-12}$ and the y-intercept is $b = \boxed{6}$.

③ First isolate y. Add $3y$ to both sides to get $3y + 6 = 9x$. Subtract 6 from both sides to get $3y = 9x - 6$. Divide both sides by 3 to get $y = 3x - 2$. Compare $y = 3x - 2$ with $y = mx + b$ to see that the slope is $m = \boxed{3}$ and the y-intercept is $b = \boxed{-2}$.

④ Look closely: The given equation has x solved for instead of y. First isolate y. Add $-2y$ to both sides to get $2y + x = 0$. Subtract x from both sides to get $2y = -x$. Divide both sides by 2 to get $y = -\frac{x}{2}$. Compare $y = -\frac{x}{2}$ with $y = mx + b$ to see that the slope is $m = \boxed{-\frac{1}{2} = -0.5}$ and the y-intercept is $b = \boxed{0}$.

Page 186

There aren't any problems on page 186.

Page 187

① $x_1 = 4, y_1 = 3, x_2 = 8, y_2 = 5$: $m = \frac{y_2 - y_1}{x_2 - x_1} = \frac{5-3}{8-4} = \frac{2}{4} = \frac{1}{2} = 0.5$

Plug $x = 4$ and $y = 3$ for $(4, 3)$ along with $m = 0.5$ into $y = mx + b$ to get

$3 = 0.5(4) + b$, which becomes $3 = 2 + b$ or $3 - 2 = 1 = b$.

Plug $m = 0.5$ and $b = 1$ into $y = mx + b$ to get $\boxed{y = 0.5x + 1}$.

② $x_1 = 3, y_1 = -4, x_2 = -2, y_2 = 6$: $m = \frac{y_2 - y_1}{x_2 - x_1} = \frac{6-(-4)}{-2-3} = \frac{6+4}{-5} = \frac{10}{-5} = -2$

Plug $x = 3$ and $y = -4$ for $(3, -4)$ along with $m = -2$ into $y = mx + b$ to get

$-4 = -2(3) + b$, which becomes $-4 = -6 + b$ or $6 - 4 = 2 = b$.

Plug $m = -2$ and $b = 2$ into $y = mx + b$ to get $\boxed{y = -2x + 2}$.

Page 188

There aren't any problems on page 188.

Page 189

① $x_1 = 4, y_1 = 1, x_2 = 8, y_2 = 3$: $m = \frac{y_2 - y_1}{x_2 - x_1} = \frac{3-1}{8-4} = \frac{2}{4} = \frac{1}{2} = 0.5$

Plug $x = 4$, $y = 1$, and $m = 0.5$ into $y = mx + b$ to get $1 = 0.5(4) + b$, which

becomes $1 = 2 + b$ or $1 - 2 = -1 = b$.

Plug $m = 0.5$ and $b = -1$ into $y = mx + b$ to get $\boxed{y = 0.5x - 1}$.

When $x = 12$, we get $y = 0.5(12) - 1 = 6 - 1 = 5$.

When $y = 7$, we get $7 = 0.5x - 1 \rightarrow 7 + 1 = 0.5x \rightarrow 8 = 0.5x \rightarrow 8(2) = 16 = x$.

x	4	8	12	16
y	1	3	5	7

② $x_1 = -4, y_1 = -9, x_2 = -2, y_2 = -3$: $m = \frac{y_2 - y_1}{x_2 - x_1} = \frac{-3-(-9)}{-2-(-4)} = \frac{-3+9}{-2+4} = \frac{6}{2} = 3$

Plug $x = -4$, $y = -9$, and $m = 3$ into $y = mx + b$ to get $-9 = 3(-4) + b$,

which becomes $-9 = -12 + b$ or $12 - 9 = 3 = b$.

Plug $m = 3$ and $b = 3$ into $y = mx + b$ to get $\boxed{y = 3x + 3}$.

When $x = 2$, we get $y = 3(2) + 3 = 6 + 3 = 9$.

When $y = 3$, we get $3 = 3x + 3 \rightarrow 3 - 3 = 3x \rightarrow 0 = 3x \rightarrow 0 = x$.

x	-4	-2	0	2
y	-9	-3	3	9

Page 190

There aren't any problems on page 190.

Page 191

① The endpoints are $(-2, -5)$ and $(3, 5)$.

Note that $x_1 = -2$, $y_1 = -5$, $x_2 = 3$, and $y_2 = 5$.

$$m = \frac{y_2 - y_1}{x_2 - x_1} = \frac{5 - (-5)}{3 - (-2)} = \frac{5 + 5}{3 + 2} = \frac{10}{5} = 2$$

The y-intercept is $b = -1$. This is where the line crosses the vertical axis.

Plug $m = 2$ and $b = -1$ into $y = mx + b$ to get $\boxed{y = 2x - 1}$.

② The endpoints are $(0, 0)$ and $(10, 2.5)$.

Note that $x_1 = 0$, $y_1 = 0$, $x_2 = 10$, and $y_2 = 2.5$.

$$m = \frac{y_2 - y_1}{x_2 - x_1} = \frac{2.5 - 0}{10 - 0} = \frac{2.5}{10} = 0.25 \text{ (alternative: } m = \frac{1}{4})$$

The y-intercept is $b = 0$. This is where the line crosses the vertical axis.

Plug $m = 0.25$ and $b = 0$ into $y = mx + b$ to get $\boxed{y = 0.25x}$. Alternative: $y = \frac{x}{4}$

Page 192

There aren't any problems on page 192.

Page 193

① $y = -2(0) + 9 = 0 + 9 = 9$

$y = -2(1) + 9 = -2 + 9 = 7$

$y = -2(2) + 9 = -4 + 9 = 5$

$y = -2(3) + 9 = -6 + 9 = 3$

x	y
0	9
1	7
2	5
3	3

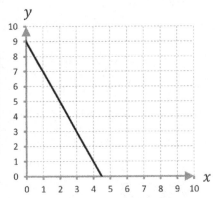

② Note: Neither $x = 0$ nor $x = 1$ will show up on the given plot because for those values of x, it turns out that y is negative. Thus, we begin with $x = 2$.

$$y = 4(2) - 8 = 8 - 8 = 0$$

$$y = 4(3) - 8 = 12 - 8 = 4$$

$$y = 4(4) - 8 = 16 - 8 = 8$$

$$y = 4(5) - 8 = 20 - 8 = 12$$

x	y
2	0
3	4
4	8
5	12

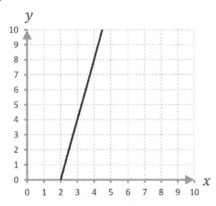

Page 194

① (C) $(3, 7)$ since it is $x = 3$ to the right and $y = 7$ above the origin.

② (B) II (compare with the diagram in Sec. 8.2)

③ (B) $(-3, 4)$ since it is 3 units to the left (which makes $x = -3$) and 4 units above (which makes $y = 7$) the origin.

Page 195

④ (A) 0.2 since $m = \dfrac{y_2 - y_1}{x_2 - x_1} = \dfrac{-4 - (-8)}{10 - (-10)} = \dfrac{-4 + 8}{10 + 10} = \dfrac{4}{20} = \dfrac{1}{5} = 0.2$

Note that $x_1 = -10$, $y_1 = -8$, $x_2 = 10$, and $y_2 = -4$.

We used the endpoints, which are $(-10, -8)$ and $(10, -4)$.

⑤ (B) -6 since this is the value of y when the line crosses the vertical axis.

⑥ (A) $y = -3x + 9$ since $m = \dfrac{y_2 - y_1}{x_2 - x_1} = \dfrac{0 - 9}{3 - 0} = \dfrac{-9}{3} = -3$ and $b = 9$

Note that $x_1 = 0$, $y_1 = 9$, $x_2 = 3$, and $y_2 = 0$.

We used the endpoints, which are $(0, 9)$ and $(3, 0)$.

Page 196

⑦ (E) 3 since $m = \frac{y_2 - y_1}{x_2 - x_1} = \frac{20 - 2}{10 - 4} = \frac{18}{6} = 3$

Note that $x_1 = 4$, $y_1 = 2$, $x_2 = 10$, and $y_2 = 20$.

⑧ (A) 0.2 (compare $y = mx + b$ with $y = 0.2x - 8$)

⑨ (D) -8 (compare $y = mx + b$ with $y = 0.2x - 8$)

⑩ (A) -3 (compare $y = mx + b$ with $y = -3x - 2$)

Subtract $9x$ and 6 from both sides of $9x + 3y + 6 = 0$ to get $3y = -9x - 6$.

Divide both sides by 3 to get $y = -3x - 2$.

⑪ (B) -2 (compare $y = mx + b$ with $y = -3x - 2$ from Problem 10)

⑫ (D) $y = 3x - 1$ since $m = \frac{y_2 - y_1}{x_2 - x_1} = \frac{14 - 8}{5 - 3} = \frac{6}{2} = 3$ and $8 = 3(3) + b$ becomes

$8 = 9 + b$ and $8 - 9 = -1 = b$ such that $y = mx + b = 3x + (-1) = \boxed{3x - 1}$

Note that $x_1 = 3$, $y_1 = 8$, $x_2 = 5$, and $y_2 = 14$.

To find b, we plugged $x = 3$ and $y = 8$ for $(3, 8)$ into $y = mx + b$ (see above).

Check: Plug $x = 3$ into $y = 3x - 1$ to get $y = 3(3) - 1 = 9 - 1 = 8$ and plug

$x = 5$ into $y = 3x - 1$ to get $y = 3(5) - 1 = 15 - 1 = 14$.

⑬ (A) $y = -2x + 1$ since $m = \frac{y_2 - y_1}{x_2 - x_1} = \frac{-1 - 5}{1 - (-2)} = \frac{-6}{1 + 2} = -\frac{6}{3} = -2$ and

$5 = -2(-2) + b$ becomes $5 = 4 + b$ and $5 - 4 = 1 = b$ such that

$y = mx + b = \boxed{-2x + 1}$

Note that $x_1 = -2$, $y_1 = 5$, $x_2 = 1$, and $y_2 = -1$.

To find b, we plugged $x = -2$ and $y = 5$ into $y = mx + b$ (see above).

Check: Plug $x = -2$ into $y = -2x + 1$ to get $y = -2(-2) + 1 = 4 + 1 = 5$

and plug $x = 1$ into $y = -2x + 1$ to get $y = -2(1) + 1 = -2 + 1 = -1$.

⑭ (B) -7 since $y = -2x + 1 = -2(4) + 1 = -8 + 1 = -7$ (we plugged

$x = 4$ into the equation $y = -2x + 1$ from Problem 13)

Chapter 9: Geometry

Page 197

There aren't any problems on page 197.

Page 198

There aren't any problems on page 198.

Page 199

① $7 \text{ cm} \times \frac{10 \text{ mm}}{1 \text{ cm}} = 7 \times 10 = \boxed{70 \text{ mm}}$

② $4 \text{ mm} \times \frac{1 \text{ cm}}{10 \text{ mm}} = \frac{4}{10} = \frac{2}{5} = \boxed{0.4 \text{ cm}}$

③ $2.5 \text{ gal.} \times \frac{4 \text{ qt.}}{1 \text{ gal.}} = 2.5 \times 4 = \boxed{10 \text{ qt.}}$

④ $375 \text{ g} \times \frac{1 \text{ kg}}{1000 \text{ g}} = \frac{375}{1000} = \boxed{0.375 \text{ kg}}$

⑤ $7.62 \text{ cm} \times \frac{1 \text{ in.}}{2.54 \text{ cm}} = \frac{7.62}{2.54} = \boxed{3 \text{ in.}}$

⑥ $3 \text{ days} \times \frac{24 \text{ hr.}}{1 \text{ day}} = 3 \times 24 = \boxed{72 \text{ hr.}}$

Page 200

There aren't any problems on page 200.

Page 201

① $80 \frac{\text{yd.}}{\text{min.}} \times \frac{3 \text{ ft.}}{1 \text{ yd.}} \times \frac{1 \text{ min.}}{60 \text{ s}} = \frac{240}{60} = \boxed{4 \text{ ft./s}}$

② $36 \frac{\text{in.}}{\text{s}} \times \frac{1 \text{ ft.}}{12 \text{ in.}} \times \frac{60 \text{ s}}{1 \text{ min.}} = \frac{36}{12} \times 60 = 3 \times 60 = \boxed{180 \text{ ft./min.}}$

③ $0.24 \text{ cm}^2 \times \left(\frac{10 \text{ mm}}{1 \text{ cm}}\right)^2 = 0.24 \text{ cm}^2 \times \frac{100 \text{ mm}^2}{1 \text{ cm}^2} = 0.24 \times 100 = \boxed{24 \text{ mm}^2}$

Note: $1 \text{ cm}^2 = (10 \text{ mm})^2 = 10 \text{ mm} \times 10 \text{ mm} = 100 \text{ mm}^2$

Note: The conversion factor gets squared, as explained on page 200.

④ $3600 \text{ mm}^3 \times \left(\frac{1 \text{ cm}}{10 \text{ mm}}\right)^3 = 3600 \text{ mm}^3 \times \frac{1 \text{ cm}^3}{1000 \text{ mm}^3} = \frac{3600}{1000} = \boxed{3.6 \text{ cm}^3}$

Note: $1 \text{ cm}^3 = (10 \text{ mm})^3 = 10 \text{ mm} \times 10 \text{ mm} \times 10 \text{ mm} = 1000 \text{ mm}^3$

Note: The conversion factor gets cubed, as explained on page 200.

Page 202

$$① \ 120° \times \frac{\pi \ \text{rad}}{180°} = \frac{120\pi}{180} = \boxed{\frac{2\pi}{3} \ \text{rad}}$$

$$② \ \frac{\pi}{6}\text{rad} \times \frac{180°}{\pi \ \text{rad}} = \frac{180\pi}{6\pi} = \frac{180}{6} = \boxed{30°}$$

$$③ \ \frac{3\pi}{4}\text{rad} \times \frac{180°}{\pi \ \text{rad}} = \frac{540\pi}{4\pi} = \frac{540}{4} = \boxed{135°}$$

Page 203

There aren't any problems on page 203.

Page 204

① Complementary angles: $\angle 1 + 65° = 90° \rightarrow \angle 1 = 90° - 65° = \boxed{25°}$

② Supplementary angles: $130° + \angle 2 = 180° \rightarrow \angle 2 = 180° - 130° = \boxed{50°}$

③ Vertical angles: $\angle 3 = \boxed{27°}$

④ Vertical angles: $\angle 4 = \boxed{55°}$

Complementary angles: $\angle 4 + \angle 5 = 90° \rightarrow \angle 5 = 90° - 55° = \boxed{35°}$

Page 205

① right (top has right angle symbol) and isosceles (2 sides have length 10)

② obtuse (one angle exceeds 90°) and scalene (no sides are congruent)

③ acute (all angles less than 90°) and equilateral (3 sides have length 8)

Page 206

There aren't any problems on page 206.

Page 207

① $\angle 1 + 54° + 45° = 180° \rightarrow \angle 1 + 99° = 180° \rightarrow \angle 1 = 180° - 99° = \boxed{81°}$

② $\angle 2 + 90° + 50° = 180° \rightarrow \angle 2 + 140° = 180° \rightarrow \angle 2 = 180° - 140° = \boxed{40°}$

Note: The □ symbol indicates that this is a right triangle. It has one 90° angle.

③ Supplementary angles: $\angle 3 + 140° = 180° \rightarrow \angle 3 = 180° - 140° = \boxed{40°}$

$\angle 3 + \angle 4 + \angle 4 = 180° \rightarrow 40° + 2(\angle 4) = 180° \rightarrow 2(\angle 4) = 180° - 40°$

$2(\angle 4) = 140° \rightarrow \angle 4 = \frac{140°}{2} = \boxed{70°}$

④ $88° + \angle 6 + \angle 6 = 180° \rightarrow 2(\angle 6) = 180° - 88° \rightarrow 2(\angle 6) = 92°$ (divide by 2)

$\angle 6 = \frac{92°}{2} = \boxed{46°}$

$34° + 63° + \angle 5 = 180° \rightarrow \angle 5 + 97° = 180° \rightarrow \angle 5 = 180° - 97° = \boxed{83°}$

Page 208

There aren't any problems on page 208.

Page 209

① possible

② impossible because $6 + 1 = 7$ isn't greater than 8

③ impossible because $11 + 18 = 29$ isn't greater than 30

④ possible

⑤ impossible because $2.3 + 0.9 = 3.2$ isn't greater than 3.5

Page 210

There aren't any problems on page 210.

Page 211

① $9^2 + 12^2 = w^2 \rightarrow 81 + 144 = w^2 \rightarrow 225 = w^2 \rightarrow \sqrt{225} = \sqrt{w^2} \rightarrow \boxed{15} = w$

② $x^2 + 10^2 = 26^2 \rightarrow x^2 + 100 = 676 \rightarrow x^2 = 576 \rightarrow \sqrt{x^2} = \sqrt{576} \rightarrow x = \boxed{24}$

③ $4^2 + y^2 = 5^2 \rightarrow 16 + y^2 = 25 \rightarrow y^2 = 9 \rightarrow \sqrt{y^2} = \sqrt{9} \rightarrow y = \boxed{3}$

④ $8^2 + 15^2 = z^2 \rightarrow 64 + 225 = z^2 \rightarrow 289 = z^2 \rightarrow \sqrt{289} = \sqrt{z^2} \rightarrow \boxed{17} = z$

Page 212

There aren't any problems on page 212.

Page 213

①   ②

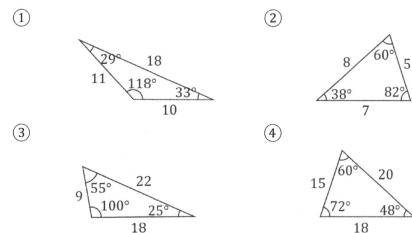

③   ④

Page 214

① rectangle   ② rhombus   ③ square   ④ trapezoid

Page 215

① $P = 6 + 6 + 3 = 12 + 3 = \boxed{15}$

Note: Two sides equal 6 since the triangle is isosceles.

② $P = 9 + 6 + 9 + 12 = \boxed{36}$

③ $P = 9 + 15 + 9 + 15 = 2(9) + 2(15) = 18 + 30 = \boxed{48}$

Note: A rectangle has two pairs of equal edge lengths.

④ $P = 18 + 18 + 18 + 18 = 4(18) = \boxed{72}$

Note: A rhombus has four congruent edges.

Page 216

① $120° = 120° \times \frac{\pi \text{ rad}}{180°} = \frac{2\pi}{3}$ rad

$s = R\theta = 9\left(\frac{2\pi}{3}\right) = \frac{18\pi}{3} = \boxed{6\pi} \approx 6(3.14) \approx \boxed{18.84}$

② $30° = 30° \times \frac{\pi \text{ rad}}{180°} = \frac{\pi}{6}$ rad

$s = R\theta = 12\left(\frac{\pi}{6}\right) = \boxed{2\pi} \approx 2(3.14) \approx \boxed{6.28}$

③ $360° = 360° \times \frac{\pi \text{ rad}}{180°} = 2\pi$ rad

$s = R\theta = 3(2\pi) = \boxed{6\pi} \approx 6(3.14) \approx \boxed{18.84}$

Note: The angle corresponding to a full circle is 360° (Sec. 9.2).

Page 217

There aren't any problems on page 217.

Page 218

There aren't any problems on page 218.

Page 219

① $A = \frac{1}{2}bh = \frac{1}{2}(11)(6) = \frac{66}{2} = \boxed{33}$ cm²

② First find the radius: $R = \frac{D}{2} = \frac{24}{2} = 12$ m

$A = \pi R^2 = \pi(12)^2 = \boxed{144\pi}$ m² $\approx 144(3.14) \approx \boxed{452.16}$ m²

③ $A = \frac{1}{2}(b_1 + b_2)h = \frac{1}{2}(3 + 5)(3) = \frac{1}{2}(8)(3) = \frac{24}{2} = \boxed{12}$ ft.²

④ $A = \frac{1}{2}d_1 d_2 = \frac{1}{2}(25)(40) = \frac{1000}{2} = \boxed{500}$ mm²

Page 220

    ① $V = LWH = (9)(5)(4) = \boxed{180}$ cm³

Page 221

    ① $A = \frac{1}{2}bh \rightarrow 24 = \frac{1}{2}(6)h \rightarrow 24 = 3h \rightarrow \frac{24}{3} = \boxed{8}$ in. $= h$

    ② $A = L^2 \rightarrow 9 = L^2 \rightarrow \sqrt{9} = \sqrt{L^2} \rightarrow \boxed{3}$ ft. $= L$

    ③ $V = L^3 \rightarrow 64 = L^3 \rightarrow \sqrt[3]{64} = \sqrt[3]{L^3} \rightarrow \boxed{4}$ cm $= L$

    Note: $\sqrt[3]{64} = 4$ (called the cube root) since $4^3 = 4 \times 4 \times 4 = 64$.

Page 222

    ① (C) 6.5 ft. since 78 i̶n̶.$\times \frac{1 \text{ ft.}}{12 \text{ i̶n̶.}} = \frac{78}{12} = \frac{78 \div 6}{12 \div 6} = \frac{13}{2} = \frac{13 \times 5}{2 \times 5} = \frac{65}{10} = 6.5$ ft.

    Note: Although 78 inches equates to 6 feet plus 6 inches, it doesn't equal 6.6 feet because 6 inches equals 0.5 feet (not 0.6 feet). The correct answer is 6.5.

    ② (D) 162 ft.² since 1 yd.² = 3 ft. × 3 ft. = 9 ft.² (as explained in Sec. 9.1)

Thus, 18 yd.² $\times \frac{9 \text{ ft.}^2}{1 \text{ yd.}^2} = 162$ ft.²

    ③ (B) 5.4 km/hr. since 1.5 $\frac{\text{m̶}}{\text{s̶}} \times \frac{1 \text{ km}}{1000 \text{ m̶}} \times \frac{3600 \text{ s̶}}{1 \text{ hr.}} = 5.4$ km/hr.

    ④ (D) $\frac{3\pi}{2}$ rad since 270° $\times \frac{\pi \text{ rad}}{180°} = \frac{3\pi}{2}$ rad

    ⑤ (D) obtuse since the angle opposite to 22 is greater than 90°

    ⑥ (C) 12 since $5 + 8 > 12$, $5 + 12 > 8$, and $8 + 12 > 5$

Page 223

    ⑦ (A) 27° since ∠1 and 63° are complements:

∠1 + 63° = 90° → ∠1 = 90° − 63° = 27°

    ⑧ (D) 117° since ∠2 and 63° are supplements:

∠2 + 63° = 180° → ∠2 = 180° − 63° = 117°

    ⑨ (C) 63° since ∠3 and 63° are vertical angles (so they are congruent)

    ⑩ (A) $3 < x < 15$ since $6 + x > 9$ leads to $x > 9 - 6 \rightarrow x > 3$ and since

$6 + 9 > x$ leads to $15 > x$. Combine $x > 3$ and $15 > x$ to get $3 < x < 15$.

    ⑪ (C) 67° since $42° + 71° + x = 180°$ according to the angle sum theorem,

such that $x = 180° - 42° - 71° = 180° - 113° = 67°$

    ⑫ (D) rhombus (four congruent sides)

Page 224

⑬ (C) 1 since $0.6^2 + 0.8^2 = x^2$ according to the Pythagorean theorem, such that $0.36 + 0.64 = x^2 \rightarrow 1 = x^2 \rightarrow \sqrt{1} = \sqrt{x^2} \rightarrow 1 = x$

⑭ (D) 250 in.^{2} since $A = \frac{1}{2}bh = \frac{1}{2}(20)(25) = \frac{500}{2} = 250$ in.^{2}

⑮ (B) 31 m since $P = 8 + 7 + 6 + 10 = 31$ m

Note: Question 15 asks for perimeter, not area. (For area, see Question 16.)

⑯ (D) 51 m^2 since $A = \frac{1}{2}(b_1 + b_2)h = \frac{1}{2}(7 + 10)(6) = \frac{1}{2}(17)(6) = \frac{102}{2} = 51$ m^2

⑰ (C) 0.6 cm since $A = L^2$, such that $0.36 = L^2 \rightarrow \sqrt{0.36} = \sqrt{L^2} \rightarrow 0.6 = L$

Note: $\sqrt{0.36} = 0.6$ because $(0.6)^2 = 0.6 \times 0.6 = 0.36$

⑱ (B) 5 ft. since $V = LWH$ and $A = LW$ such that $V = AH$ and $60 = 12H$:

$\frac{60}{12} = 5 = H$

Chapter 10: Finance

Page 225

There aren't any problems on page 225.

Page 226

(1) $20(25) + 8(10) + 9(5) + 29(1) = 500 + 80 + 45 + 29 = 654¢$

$654¢ \times \frac{\$1}{100¢} = \boxed{\$6.54}$

(2) 3 quarters, 1 dime, 1 nickel, 2 pennies

$3(25) + 1(10) + 1(5) + 2(1) = 75 + 10 + 5 + 2 = 92¢$ and $92¢ \times \frac{\$1}{100¢} = \boxed{\$0.92}$

Page 227

(1) $(\$0.05)x = \$1.2 \rightarrow 5x = 120 \rightarrow x = \frac{120}{5} = \boxed{24}$ Check: $(\$0.05)24 = \$1.2$

(2) $\$2.85 + x = \$5.15 \rightarrow 285 + 100x = 515 \rightarrow 100x = 515 - 285$

$100x = 230 \rightarrow x = \frac{230}{100} = \boxed{\$2.30}$ Check: $\$2.85 + \$2.30 = \$5.15$

Page 228

(1) $\$25 \times \frac{£1}{\$1.25} = \frac{25 \times 4}{1.25 \times 4} = \frac{100}{5} = \boxed{£20}$

(2) $12€ \times \frac{\$1.1}{1€} = \boxed{\$13.20}$

Notes: The euro € symbol usually appears after a value rather than in front of it, unlike many other units of currency. Also, note that it would be unusual to write $13.2 instead of $13.20. If writing a decimal value for cents, it is usually expressed in hundredths (not tenths), since one cent equals $0.01.

(3) $¥80 \times \frac{\$0.009}{1¥} = \boxed{\$0.72}$

(4) $\$30 \times \frac{\text{Rs } 1}{\$0.015} = \frac{30 \times 200}{0.015 \times 200} = \frac{6000}{3} = \boxed{\text{Rs } 2000}$

Page 229

(1) Convert the tax rate to a decimal: $9\% = \frac{9\%}{100\%} = 0.09$.

The tax is $\$11 \times 0.09 = \$0.99$. The total cost is $\$11 + \$0.99 = \boxed{\$11.99}$.

② Convert the tax rates to decimals: $4\% = \frac{4\%}{100\%} = 0.04$ and $10\% = \frac{10\%}{100\%} = 0.1$.

The tax is $\$4 \times 0.04 = \$0.16$ for the milk and $\$12 \times 0.1 = \$1.20$ for the shirt.

The total cost is $\$4 + \$0.16 + \$12 + \$1.20 = \boxed{\$17.36}$.

Page 230

① Convert the interest rate to a decimal: $2\% = \frac{2\%}{100\%} = 0.02$.

The interest earned is $I = Pr = \$120 \times 0.02 = \$2.40$.

Note: Unlike the example (and Problem 2), the first problem didn't ask for the total amount. It only asked how much interest would be earned.

② Convert the interest rate to a decimal: $7\% = \frac{7\%}{100\%} = 0.07$.

The interest earned is $I = Pr = \$200 \times 0.07 = \$14$.

The total amount is $\$200 + \$14 = \boxed{\$214}$.

Page 231

There aren't any problems on page 231.

Page 232

There aren't any problems on page 232.

Page 233

① Since $\frac{\$10}{\$0.50} = \frac{10 \times 2}{0.5 \times 2} = \frac{20}{1} = 20$, you would need to use your debit card more than 20 times per month in order to make it worthwhile to pay the $10 monthly fee. Therefore, Bank X makes sense if you plan to make more than 20 debit transactions per month, while Bank Y makes sense if you don't plan to make 20 debit transactions per month.

Page 234

① debit　　　② debit　　　③ credit　　　④ debit

Page 235

There aren't any problems on page 235.

Page 236

   ① Add the credits, but subtract the debits. Examples: $1832.27 - $550.00 = 1282.27 and $1282.27 + $743.85 = 2026.12.

Check Number	Date	Transaction	Credit	Debit	Balance
	7/1	beginning balance			$1832.27
517	7/1	check rent		$550.00	$1282.27
	7/3	deposit from paycheck	$743.85		$2026.12
	7/4	ATM withdrawal		$300.00	$1726.12
	7/7	deposit birthday cash	$150.00		$1876.12
518	7/9	check car payment		$362.75	$1513.37
	7/11	debit card car repairs		$261.38	$1251.99

Page 237

   ① a low debt-to-income ratio. When a person must use a larger percentage of their income to pay their debts, that person is at greater risk of not being able to keep up with their payments.

Page 238

   ① (D) $5.77 since $18(25) + 7(10) + 8(5) + 17(1) = 450 + 70 + 40 + 17 = 577$¢ and 577¢ $\times \frac{$1}{100¢} = 5.77

   ② (B) 25 since $($0.25$)x = $6.25 \rightarrow 25x = 625 \rightarrow x = \frac{625}{25} = 25$

Check: $($0.25$)25 = 6.25

   ③ (A) $2.89 since $3.29 + x = $6.18 \rightarrow 329 + 100x = 618 \rightarrow$

$100x = 618 - 329 \rightarrow 100x = 289 \rightarrow x = \frac{289}{100} = 2.89

Check: $3.29 + $2.89 = 6.18

④ (A) £24 since $\$30 \times \dfrac{£1}{\$1.25} = \dfrac{30 \times 4}{1.25 \times 4} = \dfrac{120}{5} = £24$

⑤ (E) $\$56.25$ since $£45 \times \dfrac{\$1.25}{£1} = \$56.25$

⑥ (C) $\$3.36$ since $8\% = \dfrac{8\%}{100\%} = 0.08$ and $\$42 \times 0.08 = \$3.36$

⑦ (E) $\$302.50$ since $10\% = \dfrac{10\%}{100\%} = 0.1$, $\$275 \times 0.1 = \$27.50$, and $\$275 + \$27.50 = \$302.50$

Page 239

⑧ (C) $\$10$ since $2.5\% = \dfrac{2.5\%}{100\%} = 0.025$ and $\$400 \times 0.025 = \$10$

⑨ (E) $\$2520$ since $5\% = \dfrac{5\%}{100\%} = 0.05$, $\$2400 \times 0.05 = \$120$, and $\$2400 + \$120 = \$2520$

⑩ (E) 15 since $\dfrac{\$6}{\$0.40} = \dfrac{6 \times 10}{0.4 \times 10} = \dfrac{60}{4} = 15$

Check: $15(\$0.40) = \$6$

⑪ (A) payroll deposit (this option adds to your checking account balance)

⑫ (C) $\$584.92$ since $\$632.85 - \$47.93 = \$584.92$

Page 240

⑬ (B) $\$90$ since $25\% = \dfrac{25\%}{100\%} = 0.25 = \dfrac{1}{4}$ and $\$360 \times 0.25 = \$360 \times \dfrac{1}{4} = \$90$

⑭ (D) low debt-to-income ratio

GLOSSARY

absolute value: a number's value after removing its sign. Absolute values are indicated by placing vertical lines on each side of a number. For example, $|-5| = 5$.

acute: an angle with an angular measure that is less than 90°. An acute triangle has three acute angles.

angle sum theorem: the sum of the angular measures of the interior angles of any triangle is 180°.

arc length: the distance along a section of a curve.

associative property: the result of adding or multiplying three or more numbers does not depend on how the numbers are grouped; $(a + b) + c = a + (b + c)$ and $(ab)c = a(bc)$.

base: any number that has an exponent (or power). In 3^4, for example, the base is 3.

central angle: an angle that is measured from the center of a circle.

circumference: the total distance around a circle.

coefficient: a number that multiplies a variable. In $4x$, for example, the coefficient is 4.

commutative property: the result of adding or multiplying two numbers does not depend on the order; $a + b = b + a$ and $ab = ba$.

complementary angles: two angles that form a right angle. Their angular measures add up to 90°.

congruent: two geometric figures that have both the same shape and the same size.

constant: a fixed value. For example, in $x + 7$, the number 7 is the constant.

credit: a transaction that adds to the balance of a checking account.

cube: 1) raise a number to the power of three. For example, 4^3 is four cubed. 2) a three-dimensional solid formed from six congruent square sides.

cuboid: a rectangular prism, which is a three-dimensional solid formed from six rectangles.

cylinder: a three-dimensional solid that has two congruent circles at its ends and a surface perpendicular to the circles.

debit: a transaction that reduces the balance of a checking account.

debt-to-income ratio: the ratio of a person's total debt to the person's total income.

degree: one 360th of a circle.

denominator: the number at the bottom of a fraction. In $\frac{2}{7}$, for example, the denominator is 7.

density: the ratio of an object's mass to its volume. It is a measure of the compactness of a substance.

deposit: money that is added to a bank account.

diameter: a line that passes through the center of a circle and which connects points on opposite ends of the circle. It is twice the radius.

distributive property: $a(b + c) = ab + ac$.

equilateral: a polygon for which all of the sides have the same length.

exponent: the number of times that the base is multiplied by itself. For example, in 4^3, the exponent is 3, causing the base to be multiplied by itself three times; $4^3 = 4 \times 4 \times 4$.

f.o.i.l. method: first, outside, inside, last. A method of multiplying; $(w + x)(y + z) = wy + wz + xy + xz$.

factorization: the prime numbers that multiply together to form a whole number. For example, $30 = 2 \times 3 \times 5$ is the prime factorization of 30.

factor: 1) a number that is being multiplied. For example, in $4 \times 7 = 28$, the numbers 4 and 7 are called factors. 2) the distributive property applied backwards; $ab + ac = a(b + c)$.

frequency: the number of times that a result occurs in a data set.

GCF: the greatest common factor. For example, the GCF of 16 and 24 equals 8 since $16 = 2 \times 8$ and $24 = 3 \times 8$.

grace period: the number of days (such as 25) for which a person can pay the previous month's balance on a credit card without having to pay interest (provided that interest hadn't already been accruing prior to the previous month).

horizontal: a line running across from left to right without any steepness. A horizontal line has zero slope.

hypotenuse: the longest side of a right triangle, opposite to the 90° angle.

identity property: adding zero or multiplying by one have no effect; $x + 0 = x$ and $1x = x$.

improper fraction: a fraction for which the numerator is larger than the denominator, such as $\frac{5}{4}$.

integer: whole numbers like 0, 1, 2, 3, 4, etc., and negative values like $-1, -2, -3, -4$, etc. See the definitions for whole and natural numbers regarding differences from integers.

intercept: the point where a line or curve crosses an axis.

interest: money that is earned from savings or investments, or money that is paid for taking out a loan.

interior angle: the angle between two consecutive edges of a polygon.

interquartile range: a measure of the spread of a data set equal to the difference between the upper quartile and the lower quartile.

inverse property: the opposite of adding or multiplying by a number; $x + (-x) = 0$ and $x \left(\frac{1}{x} \right) = 1$.

inverse proportion: quantities that are related by $y = \frac{c}{x}$, where c is a constant. An increase in one quantity causes a decrease in the other quantity.

investment: to put money into a savings account, business, stocks, etc., with the potential to make a profit.

IQR: interquartile range.

irrational: a real number such as $\sqrt{2}$ or π which cannot be expressed as an integer or the ratio of two integers.

isosceles: a triangle with at least two congruent edges.

LCD: lowest common denominator.

LCM: least common multiple. For example, 12 is the LCM of 4 and 6 since $4 \times 3 = 12$ and $6 \times 2 = 12$.

leaf: the units digit of a two-digit number on a stem-and-leaf plot. In 24, for example, 2 is the stem and 4 is the leaf.

legs: the two shortest sides of a right triangle.

like terms: terms in an expression where the same variable is raised to the same power, such as $3x^2$ and $5x^2$.

lower quartile: the median value of the lower half of a data set.

mean value: the average value of a data set.

median: the middle value of a data set. For a data set with an even number of values, find the mean of the two middle numbers.

mixed number: a number that includes an integer plus a fraction, such as $4\frac{2}{5}$ (which equals four plus two-fifths).

mode: the result in a data set that occurs most frequently.

mortgage loan: a loan used to purchase property.

natural number: positive numbers that are whole, like 1, 2, 3, 4, etc. Some books and instructors distinguish between natural numbers and whole numbers, the difference being that natural numbers don't include zero.

numerator: the number at the top of a fraction. In $\frac{2}{7}$, for example, the numerator is 2.

obtuse: an angle that is greater than 90°. An obtuse triangle contains one obtuse angle.

operator: a mathematical process, such as multiplication or subtraction.

ordered pair: a pair of numbers in the form (x, y) locating a point on the coordinate plane.

origin: the point $(0, 0)$ on the coordinate plane where the x- and y-axes intersect.

outlier: a data point which lies apart from most of the other points in a data set.

parallel: two lines that are equally spaced apart so as not to intersect (even if they were extended indefinitely).

parallelogram: a quadrilateral with two pairs of parallel edges.

PEMDAS: parentheses, exponents, multiplication/division left to right, addition/subtraction left to right. This is the order of arithmetic operations.

percent: a fraction of one hundred. A percent is a specific value, like 12% or 130% (unlike a percentage).

percentage: an unspecified amount, like "a percentage of the class" (unlike a percent, which is a definite value).

perfect square: a whole number that equals a smaller whole number multiplied by itself. For example, 16 is a perfect square because $4^2 = 4 \times 4 = 16$.

perimeter: the sum of the lengths of the edges of a polygon.

perpendicular: at right angles to one another.

pi: the ratio of the circumference of a circle to its diameter, which is approximately $\pi \approx 3.14159265$ for all circles.

polygon: a closed geometric figure formed from straight edges.

power: the number of times that the base is multiplied by itself. For example, in 4^3, the power is 3, causing the base to be multiplied by itself three times; $4^3 = 4 \times 4 \times 4$.

prime number: a whole number that is evenly divisible only by itself and one, such as 2, 3, 5, 7, and 11.

principal: the amount of money invested or borrowed before the interest rate is applied.

proportion: an equality between two ratios or rates (that obey a linear relationship).

Pythagorean theorem: the sum of the squares of the legs of a right triangle equals the square of the hypotenuse.

quadrant: one of the four regions formed by the x- and y-axes intersecting in the coordinate plane.

quadrilateral: a polygon with four edges.

qualitative: involving characteristics that can be described without numerical measurements, such as color.

quantitative: involving characteristics that can be counted, measured, or expressed using numerical values.

radian: a unit of angular measure related to degrees by the conversion factor π rad $= 180°$.

radius: a line connecting the center of a circle to a point on the edge of the circle. It is one-half the diameter.

range: the difference between the greatest value and the least value in a set of a data. It provides a measure of the full spread of the data.

rate: a fraction made by dividing quantities with different units, such as $\frac{200 \text{ miles}}{3 \text{ hours}}$.

ratio: a fixed relationship expressed as a fraction in the form $a{:}b$.

rational: a real number that can be expressed as an integer (like 6) or the ratio of two integers (like $\frac{2}{3}$).

real number: a number that is rational or irrational, and which does not have an imaginary part.

reciprocal: one divided by a number. For a fraction, swap the numerator and denominator to find the reciprocal. For example, $\frac{5}{3}$ is the reciprocal of $\frac{3}{5}$.

rectangle: a parallelogram with four 90° interior angles.

rectangular prism: a three-dimensional solid formed from six rectangles.

reduced fraction: the simplest form of a fraction, where the numerator and denominator do not share a GCF.

register: a book in which banking transactions are recorded in order to keep track of the account balance.

relative frequency: the ratio of a particular frequency to the sum of the frequencies.

remainder: the amount left over when a number is divided by another number. For example, $17 \div 5$ equals three with a remainder of two since $3 \times 5 = 15$ and $17 - 15 = 2$.

repeating decimal: a digit or a group of digits that repeat forever in a decimal number, such as $0.\overline{42} = 0.42424242...$

rhombus: a parallelogram with four congruent edges.

right: a 90° angle. A right triangle includes one right angle. Perpendicular lines intersect at right angles.

rise: the vertical height between two points when finding the slope of a line.

run: the horizontal base between two points when finding the slope of a line.

scalene: a triangle with no congruent edges.

simplify: make an expression simpler.

slope: a measure of the steepness of a line determined by dividing the rise by the run.

solve: determine the value(s) of the variable(s) by following a procedure (like isolating the unknown).

spread: an indication of how much the values in a data set vary, such as the range or standard deviation.

square: 1) raise a number to the power of two. For example, 3^2 is three squared. 2) a quadrilateral that has 90° angles and also has four congruent edges.

square root: a number that when multiplied by itself makes the indicated value. For example, $\sqrt{9} = 3$ since $3^2 = 3 \times 3 = 9$.

standard deviation: a measure of the spread of a data set found by averaging the squares of the deviations from the mean and then taking a square root.

stem: the tens digit of a two-digit number on a stem-and-leaf plot. In 24, for example, 2 is the stem and 4 is the leaf.

supplementary angles: two angles that form a straight line. Their angular measures add up to 180°.

term: expressions separated by plus, minus, or equal signs in an algebraic expression or equation. For example, $3x$, 8, and $5x$ are each terms in $3x + 8 = 5x$.

trailing zeroes: zeroes that come at the end of a number and which follow a decimal point, such as 1.300.

transaction: a process which results in a credit or debit to a bank account.

trapezoid: a quadrilateral with one pair of parallel edges.

triangle: a polygon with three edges.

triangle inequality: the edge lengths of a triangle satisfy $|x - y| < z < x + y$.

unit: 1) a standard value for making measurements, such as a meter, foot, or second. 2) the ones digit of a number.

unsecured loan: a loan that is not guaranteed by collateral (using a car title or the deed to property, for example).

upper quartile: the median value of the upper half of a data set.

variable: an unknown quantity like x or y.

vertex: a point where two lines intersect.

vertical: a line running up and down (perpendicular to the horizontal).

vertical angles: two congruent angles formed opposite to a vertex where two lines intersect.

volume: the amount of space that an object occupies.

whiskers: lines on a box plot connecting the least value to the lower quartile and upper quartile to the greatest value.

whole number: numbers that are whole, like 0, 1, 2, 3, 4, etc. Some books and instructors distinguish between whole numbers and integers, the difference being that integers include negative values.

withdrawal: money that is taken out of a bank account.

y-intercept: the point where a line or curve crosses the *y*-axis.

INDEX

A

absolute values 13, 29

account balance 234-236, 239

accounts 231-233

acute triangle 205

adding decimals 60, 72

adding fractions 39-40, 54, 119

addition 6, 9, 15-16, 39-40, 54, 60, 72, 107-108, 128, 130-132

additive inverse 141

additive relationships 128, 130-132, 145

algebra pyramid 88-91

all real numbers 121

angle from center 216

angle sum theorem 206-207, 223

angles 203-207

annual fee 231, 233

applying models 131-132, 136, 140, 145-146

applying proportions 96-97, 99-100, 133-140

applying rates 90-93

applying ratios 79-80

arc 216

arc length 216

area formulas 217-219, 221, 224

area units 200-201, 222

arithmetic 6-7, 9-10, 14-17, 103

arithmetic mean 148

associative 15, 141

ATM machine 231, 234-236, 239

average 148, 167

axes 174-177

B

balance an account 234-236, 239

bank accounts 231-233

bar graphs 164-165, 171

base 11, 58

box plots 155-156, 169

box-and-whisker plots 155-156, 169

British pounds 228, 238

C

car loan 231

categorical data 157-167, 172

categories 157-167, 172

CD's 231

center 150, 167, 216

central angle 216

cents 170, 226, 238

check register 235-236

checking account 231-236, 239

checks 231, 234-236, 239

circle 216, 218-219

circumference 216

classifying numbers 70

clusters 167-168

coefficients 102, 110, 124, 134

coins 170, 226-227, 238

combine like terms 109-113, 124, 142, 146

common denominator 35-37, 39, 53, 119

commutative 15, 141

comparing decimals 57, 69, 72-73

comparing fractions 37-38, 53, 69, 72-73

comparing numbers 13, 29-30, 37-38, 57, 69, 72-73

comparing rational numbers 69, 73

complementary angles 203-204, 223

computations 17, 51, 71, 106

congruent 205, 214

constant speed 86-88, 99-100

constants 102, 109-110

conversions 198-202, 222, 228 (also see rates)

coordinate graphs 173-196

coordinate plane 174

coordinates 174-179, 186-189, 194, 196

corresponding parts 212-213

credit card 231-234, 240

credit history 232, 237

credit report 237

credit score 237, 240

credits 234-236, 239

cross multiplying 120, 125

cube 220-221

cubed 11

cubed units 200-201

cuboid 220, 224

currency 228, 238

cylinder 220

D

data analysis 147-172

data interpretation 167-168

data point 174

debit card 231-236, 239

debits 234-236, 239

debt-to-income ratio 237, 240

decimal places 60-61, 72

decimal point 56, 59-60

decimal to fraction 64-65, 73, 92

decimal to percent 63, 73-74

decimals 55-74, 85, 227, 229-230,
238-240

degrees 202, 216, 222

denominator 32, 35, 77, 119

density 90

deposits 234-236, 239

diagonals 218-219

diameter 218

digits 56, 59, 66

dimes 170, 226, 238

direct deposit 231

direct proportions 94-97, 99-100,
133-137

distance 86-88, 99-100

distributive 15, 18, 25, 30, 48-49,
115-118, 125, 141-142, 146

dividing decimals 68, 72

dividing fractions 42-43, 50, 54

divisibility 19-20, 30

division 7, 10, 14-16, 19-20, 34,
42-43, 50, 52, 54, 68, 72, 103,
107-108

dollars 226-228, 238

dot 14

dot plots 158, 162-163, 168, 170

down payment 231

E

ellipsis 148

equal sign 104, 109

equation for a straight line
184-193, 195-196

equations 104, 107-108, 110-113,
124, 143-144

equilateral 205-206

equivalent expressions 109,
114-118, 142, 146

equivalent fractions 33-34, 37, 53,
73, 78-79, 84, 92

euro 228

even numbers 19

evenly divisible 19-20, 26, 30

exchange rates 228, 238

exponents 11, 16, 29, 44-45, 54, 58, 72, 114

expressions 104, 109, 124

F

f.o.i.l. 116, 125

factor tree 21-22

factoring 25, 49, 109, 117-118, 121, 125, 146

factorization 21-23, 30

factors 21, 23-24, 30

fees 231-233, 239

fill in the numbers 78, 84, 128-129, 133, 138, 145

finance 225-240

foil 116

foreign exchange rates 228, 238

formula for a straight line 184-193, 195-196

formula for area 217-219, 221, 224

formula for slope 134, 180-181

formula for volume 220, 221

formulas, using 106, 124, 132, 136, 140, 145-146

fraction to decimal 65, 67, 73-74, 85

fraction to ratio 76, 98

fractional powers 114

fractions 31-54, 64-67, 69-70, 72-74, 76-93, 112-113, 162

frequency 157-160, 162-167, 170-172

frequency table 160, 162, 170

full spread 150

G

GCF 24-25, 32, 49, 77, 117-118

geometry 197-224

grace period 231-232

graphing 173-196

graphing an equation 192-193

greater than 13, 37-38, 57, 69, 122-123, 126

greatest common factor 24-25, 30, 32, 49, 77, 117-118

greatest value 150, 155

GV 155

H

histograms 159-160, 171

home loan 231

horizontal 174, 178

hundredths place 56, 59

hypotenuse 210

I

ideas 105

identity 15, 141

improper fraction 33-34, 52-53

inequality 13, 37-38, 57, 69, 104,
     122-123, 126, 208-209,
     222-223

integers 70

intercept 182-191, 195-196

interest 90, 230-233

interest rate 230, 238

interior angles 206

interpolation error 180

interpret data 167-168

interquartile range 153-156, 167

intervals 159, 171

inverse property 141

inverse proportions 137-140

inverse relationships 138-140, 146

investment 230, 239

IQR 153, 155, 169

irrational 28, 46, 70, 73-74

isolate the unknown 88-90, 96-97,
     110-113, 119, 122-126, 136,
     227

isosceles 205

L

ladder diagram 23

late payment 240

LCD 35-37, 39, 53

LCM 26, 35, 37

least common multiple 26, 30, 35,
     37

least value 150, 155

leaves 161

legs 210

less than 13, 37-38, 57, 69,
     122-123, 126

like terms 109-110

line of credit 231

linear proportions 94-97, 99-100, 133-136

linear relationships 133-136, 145-146, 173-196

loans 231-232

lower half 153

lower quartile 153, 155

lowest common denominator 35-37, 39, 53

LQ 153, 155

LV 155

M

making predictions 132, 136, 140, 145-146, 196

map 100

mass 90

MD 155

mean value 148-152, 167-168, 169-171

median 149-150, 152-155, 167-171

midde value 149

minimum payments 232

mixed number 33-34, 52-53

mode 162-163, 165, 167, 170-172

modeling relationships 131-132, 134-135, 139, 143-146

money 170, 225-240

mortgate loan 231

multi-digit numbers 18, 161

multiple choice 29-30, 53-54

multiple solutions 121

multiples 19, 26

multiplication 7, 10, 14-16, 18, 42-43, 54, 60-61, 72, 102-103, 107-108, 130-132

multiplicative inverse 141

multiplicative relationships 130-132, 145

multiplying decimals 61-62, 72

multiplying fractions 42-43, 54

N

natural numbers 70

negative exponents 45, 54, 58, 72, 114

negative information 237

negative numbers 8-10, 13, 29, 70, 112-113, 122, 126, 176-177

negative slopes 178

nickels 170, 226-227, 238

ninety degrees 203

no solution 121

number line 8, 29, 155, 158, 174-177

numerator 32, 77

O

obtuse triangle 205, 222

operators 6-7, 14-17, 103, 141

order decimals 57, 69, 72-73

order fractions 53, 69, 72-73

order numbers 13, 29-30, 53, 57, 69, 72-73

order of operations 16-17, 30, 51

order rational numbers 69, 73

ordered pairs 174-177, 186-189, 194, 196

origin 174

outliers 167-168

over-the-limit fee 231

P

parallel edges 214

parallelogram 214, 218-219, 221, 223

parentheses 14, 16

part vs. whole 81-82

patterns 78, 84, 128-129, 133, 138, 145

pay over time 233

paycheck 235-236, 239

payroll deposit 235-236, 239

peaks 167-168

PEMDAS 16

pennies 170, 226, 238

percent bar graphs 164-165, 171

percents 63, 69, 73-74, 162-166, 229-230, 238-240

perfect squares 28

perimeter 215, 224

perpendicular 203

pi 202

pie charts 166

place value 56-57, 72

plotting data 155-156, 158-161,
    164-166, 171-196

plugging in 106, 124, 132, 136,
    140, 145-146

polygons 205-215, 217-219,
    221-224

positive slope 178

pounds 228, 238

powers 11, 27, 29, 58, 72, 114

predictions 132, 136, 140, 145-146,
    196

prime factorization 21-23, 30

prime numbers 21

principal 90, 230, 239

properties 15, 48, 115, 141-142,
    146

proportions 94-97, 99-100, 133-140

pyramid 88-91

Pythagorean theorem 210-211, 224

Q

quadrants 176-177, 194

quadrilaterals 214-215, 217-219,
    221, 223-224

qualitative 167-168, 172

quantitative 167

quarters 170, 226-227, 238

R

radians 202, 216, 222

radius 216, 220

range 150-156, 167-171

rate equation 86-90

rate of interest 90

rate pyramid 88-91

rate tables 84

rate to decimal 85

rate to fraction 83

rates 83-93, 98-100, 200-201, 222

ratio table 78

ratio to fraction 76, 98

rational 28, 46, 70, 73-74

rationalize 46, 54

ratios 76-82, 94-100

real numbers 121

reciprocals 41-42, 45, 53, 119

rectangle 214-215, 217-219

rectangular prism 220, 224

reduced fractions 32, 77

reduced ratios 77

register 235-236

relationship tables 128-131,
    133-135, 138-139

relationships 127-146

relative frequency 162-166, 167,
    170-172

remainders 34, 52

repeating decimals 66-67, 70

rhombus 214-215, 218-219, 223

right angle 203

right triangle 205, 210-211, 217

rise 91, 178, 180

roots 27, 46, 54, 118, 125, 151

rounding 12, 59, 72

run 91, 178, 180

rupees 228

S

sales tax 229, 238

savings account 230-231, 235,
    239-240

scale 100

scalene 205

shape 167-168

shapes 205-219, 221-224

signs 8-10, 13, 29, 112-113

simple interest 230, 239

simple models 131-132, 145-146

simplest equations 88-90, 107-108,
    124-126

simplest form 32, 77

simplify 104, 109, 114-118,
    124-125

slash 14

slope 91, 134, 178-181, 184-191,
    195-196

slope formula 134, 180-181

solve 104, 107-108, 110-113,
    119-121, 125, 143-144, 227

special angles 203-204

special solutions 121

speed 86-88, 99-100

sphere 220

spread 150-151, 153, 167

square 214, 218, 221, 224

square roots 27-28, 46, 54, 118,
    125, 151

square units 200-201, 222

squared 11, 27-28, 210-211

standard deviation 151-152, 167

steepness 178

stem-and-leaf plots 161, 171

stems 161

straight line 134, 146, 173-196

subtracting decimals 60, 72

subtracting fractions 39-40, 54, 119

subtraction 6, 9, 15-16, 39-40, 54, 60, 72, 107-108

supplementary angles 203-204, 223

symmetry 167

T

table of coordinates 188-189

table of frequencies 160, 162, 170

table of properties 141

table of rates 84

table of ratios 78

table of relationships 128-131, 133-135, 138-139

tax 229

temperature 29

ten 58

tens digit 12, 56, 161

tenths place 56, 59

terms 109-110

the part and the whole 81-82

the simplest equations 88-90, 107-108, 124-126

thousandths place 56, 59, 72

three dots 148

time 86-88, 99-100

times symbol 14, 102

trailing zeroes 57, 60-61

transactions 235-236

trapezoid 214-215, 217-219, 224

triangle inequality 208-209, 222-223

triangles 205-213, 215, 217-219, 221-224

U

unit conversions 198-202, 222, 228 (also see rates)

units cubed 200-201

units digit 12, 56, 161

units of measurement 83-93, 98-100, 198-202

units squared 200-201, 222

unknown 14, 94-96, 99, 102

unsecured loan 231

upper half 153

upper quartile 153, 155

UQ 153, 155

using formulas 106, 124, 132, 136, 140, 145-146

variable in a denominator 119, 125

variable vs. constant 102, 109-110

variables 14, 101-126

vertex 203

vertical 174, 182-183

vertical angles 203-204, 223

volume 90, 220

volume units 200-201

whiskers 155

whole numbers 28, 37, 70

whole vs. part 81-82

withdrawals 231, 234-236, 239

word problems 29-30, 53-54, 72-74, 80, 82, 86-87, 89, 91-93, 96-100, 105, 124-126, 143-146, 168-172, 194-196, 220-224, 238-240

words and variables 105, 124, 143-144, 227

writing checks 231, 234-236, 239

x- and y-coordinates 174-177, 186-189, 194, 196

yen 228

y-intercept 182-191, 195-196

zero slope 178

WAS THIS BOOK HELPFUL?

A great deal of effort and thought was put into this book, such as:

- Which topics to cover to meet the needs of students of diverse levels and abilities. Hopefully, the exercises varied from easy to challenging.
- Careful selection of problems for their instructional value.
- Including the answer to every problem, along with many solutions and explanations.
- Examples to help illustrate how to solve the problems.

If you appreciate the effort that went into making this book possible, there is a simple way that you could show it:

Please take a moment to post an honest review.

For example, you can review this book at Amazon.com or Goodreads.

Even a short review can be helpful and will be much appreciated. If you're not sure what to write, following are a few ideas, though it's best to describe what's important to you.

- Are you satisfied with the topics that were covered?
- Did you enjoy the selection of problems?
- Were you able to understand the examples and answer key?
- How much did you learn from reading and using this workbook?
- Would you recommend this book to others? If so, why?

Do you believe that you found a mistake? Please email the author, Chris McMullen, at greekphysics@yahoo.com to ask about it. One of two things will happen:

- You might discover that it wasn't a mistake after all and learn why.
- You might be right, in which case the author will be grateful and future readers will benefit from the correction. Everyone is human.

ABOUT THE AUTHOR

Dr. Chris McMullen has over 20 years of experience teaching university physics in California, Oklahoma, Pennsylvania, and Louisiana. Dr. McMullen is also an author of math and science workbooks. Whether in the classroom or as a writer, Dr. McMullen loves sharing knowledge and the art of motivating and engaging students.

The author earned his Ph.D. in phenomenological high-energy physics (particle physics) from Oklahoma State University in 2002. Originally from California, Chris McMullen earned his Master's degree from California State University, Northridge, where his thesis was in the field of electron spin resonance.

As a physics teacher, Dr. McMullen observed that many students lack fluency in fundamental math skills. In an effort to help students of all ages and levels master basic math skills, he published a series of math workbooks on arithmetic, fractions, long division, algebra, geometry, trigonometry, and calculus entitled *Improve Your Math Fluency*. Dr. McMullen has also published a variety of science books, including astronomy, chemistry, and physics workbooks.

Author, Chris McMullen, Ph.D.

ARITHMETIC

For students who could benefit from additional arithmetic practice:

- Addition, subtraction, multiplication, and division facts
- Multi-digit addition and subtraction
- Addition and subtraction applied to clocks
- Multiplication with 10-20
- Multi-digit multiplication
- Long division with remainders
- Fractions
- Mixed fractions
- Decimals
- Fractions, decimals, and percentages

www.improveyourmathfluency.com

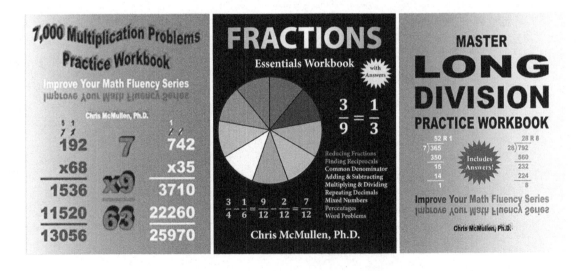

MATH

This series of math workbooks is geared toward practicing essential math skills:

- Algebra
- Geometry
- Trigonometry
- Calculus
- Fractions, decimals, and percentages
- Long division
- Multiplication and division
- Addition and subtraction

www.improveyourmathfluency.com

ALGEBRA

For students who need to improve their algebra skills:

- Isolating the unknown
- Quadratic equations
- Factoring
- Cross multiplying
- Systems of equations
- Straight line graphs
- Word problems

www.improveyourmathfluency.com

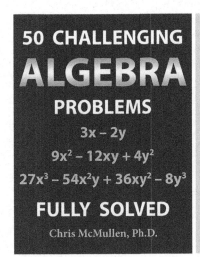

PUZZLES

The author of this book, Chris McMullen, enjoys solving puzzles. His favorite puzzle is Kakuro (kind of like a cross between crossword puzzles and Sudoku). He once taught a three-week summer course on puzzles. If you enjoy mathematical pattern puzzles, you might appreciate:

300+ Mathematical Pattern Puzzles

Number Pattern Recognition & Reasoning
- Pattern recognition
- Visual discrimination
- Analytical skills
- Logic and reasoning
- Analogies
- Mathematics

SCIENCE

Dr. McMullen has published a variety of **science** books, including:

- Basic astronomy concepts
- Basic chemistry concepts
- Balancing chemical reactions
- Calculus-based physics textbooks
- Calculus-based physics workbooks
- Calculus-based physics examples
- Trig-based physics workbooks
- Trig-based physics examples
- Creative physics problems
- Modern physics

www.monkeyphysicsblog.wordpress.com

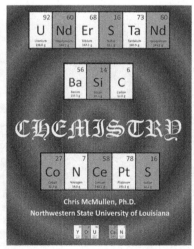

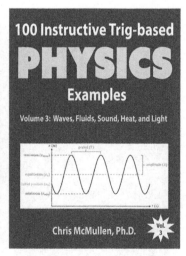

Made in the USA
Coppell, TX
06 August 2020